Zillions of Practice Problems

Pre-Algebra 0 with Physics

Zillions of Practice Problems

Pre-Algebra 0 with Physics

Stanley F. Schmidt, Ph.D.

Polka Dot Publishing

© 2017 Stanley F. Schmidt
All rights reserved.

ISBN: 978-1-937032-58-6

Printed and bound in the United States of America

Polka Dot Publishing   Reno, Nevada

To order copies of books in the Life of Fred series,

visit our website PolkaDotPublishing.com

Questions or comments? Email the author at lifeoffred@yahoo.com

First printing

Zillions of Practice Problems Pre-Algebra 0 with Physics was illustrated by the author with additional clip art furnished under license from Nova Development Corporation, which holds the copyright to that art.

for Goodness' sake

or as J.S. Bach—who was never noted for his plain English—often expressed it:

Ad Majorem Dei Gloriam
(to the greater glory of God)

If you happen to spot an error that the author, the publisher, and the printer missed, please let us know with an email to: **lifeoffred@yahoo.com**

As a reward, we'll email back to you a list of all the corrections that readers have reported.

How This Book Is Organized

Life of Fred: Pre-Algebra 0 with Physics has 37 chapters before the final bridge. So does this book.

As you work through each chapter in *Life of Fred: Pre-Algebra 0 with Physics* you can do the problems in the corresponding chapter in this book.

Each chapter in this book is divided into two parts.

★ The first part takes each topic and offers a zillion problems.

★ The second part is called the Mixed Bag. It consists of a variety of problems from the chapter and review problems from the beginning of the book up to that point.

Please write down your answers before turning to the back of the book to look at my answers. If you just read the questions and then read my answers you will learn very little. As my mother used to tell me,

No one ever climbed a hill by just looking at it.

Chapter One
Friction

First part: Problems from this chapter

104. Which is larger, a thousand or a million?

208. Which is the larger numeral, 6 or 8?

311. Here are the dimensions of the safe. What is the area of the front of the safe?

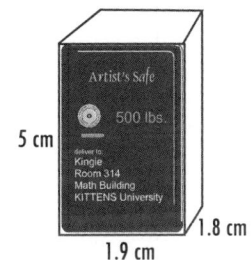

422.  They are always inventing new things.

You take the ball, roll it, and try to knock down the pins.
What new thing could they do with bowling?

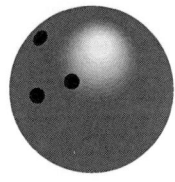

Here is a nice round ball.

Suppose I want to slide it rather than roll it.

 Chop!    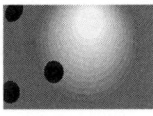 Chop!

It's now in the shape of a stick of butter.
Your question: Will it slide faster on the short side or the long side?

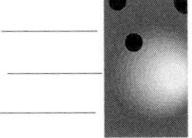   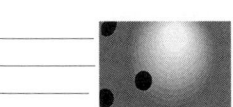

9

Chapter Two
Proportional

First part: Problems from this chapter

155. You are walking at a constant speed. The distance you walk is proportional to which of these?
    A) the number of jelly beans you have in your pocket
    B) the number of minutes you walk
    C) the day of the week
    D) the number of books you read last year

365. In physics, the letter c stands for the speed of what?

404. Some people have a lot of ducks in their living room. Some people have few ducks in their living room. About seven billion people (7,000,000,000) have zero ducks in their living room. If we let x equal the number of ducks in a person's living room, is x a discrete or a continuous variable?

560. Your pet duck is growing up. It's time to get him some shoes.
    At the store they measure his feet and claim that he needs a size 7 shoe.
    He says, "They feel a little tight."
    They give him a 7½ size.
    The duck says, "They are perfect."

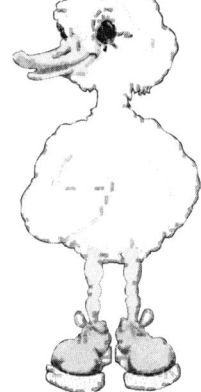

    The shoe sizes for ducks are 5, 5½, 6, 6½, 7, 7½, 8, 8½, 9, 9½, 10, 10½, and 11. Is this a discrete or continuous variable?

677. You have gone duck-nuts. The only thing you buy with your money are more ducks. Is the number of ducks you buy proportional to the amount of money you have?

Chapter Two   Proportional

Second part: the 𝕸ixed 𝕭ag: a variety of problems from this chapter and previous chapters

494. Consider your average ice-skating duck. If you are pushing him, does it take more effort if he has one foot in the air or if he has both feet on the ice?

640. If it takes 25 pounds of force to keep the duck moving at 5 mph (miles per hour), how many pounds of force will it take to move him at a constant 15 mph?

715. You never see a duck go ice skating on a carpet. It's just too much work. That illustrates the fact that friction is dependent on which of these?
    A) the area of contact between the surfaces or
    B) the speed at which he's skating or
    C) which surfaces are involved (in this case, between the bottom of his skates and the carpet)

840. Our duck decides to go out ice skating on the big lake. Is the distance he skates a discrete or a continuous variable?

905. You order a crate of ducks. Two days later the 40-pound box of ducks is sitting on your living room floor. It takes 28 pounds of force to slide it into the dining room.

Next week a 120-pound box of ducks arrives. How much force will be needed to slide it from the living room into the dining room?

11

Chapter Three
One Meter

First part: Problems from this chapter

181. Is the length of a meter in 1668 the same as the length of a meter today?

251. You want to get from one cliff to the other. The easiest way is to attach a rope to a blimp that is stationed right above the two cliffs and swing from one cliff to the other.

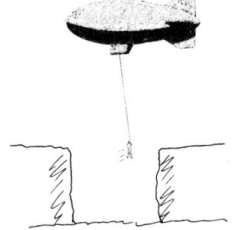

Swinging from the left cliff to the right took 12 seconds (and was a lot of fun).

Your picture appeared in the newspaper. They called you the Human Pendulum. The article said that the period was 12 seconds. Was that correct?

597. You are now standing on the right cliff and you decide to swing back to the left cliff. Your friend Jane wants to go with you. She hangs on to you, and you both make the trip together. Will it take longer or shorter than 12 seconds to get to the left cliff?

755. We know that the period of a pendulum is proportional to the square root of the length. The period is proportional to $\sqrt{\text{length}}$.
    That means that if I make the rope four times as long, the period will be twice as long. $\sqrt{4} = 2$.
    How long should the rope be in order to make the period three times as long?

926. If we make the rope one hundred times as long, how will that affect the period?

Chapter Three   One Meter

Second part: the 𝔐ixed 𝔅ag: a variety of problems from this chapter and previous chapters

280. [not an easy question] In 1668 John Wilkins defined a meter as the length of a pendulum with a period of two seconds. The French Academy (in 1791) didn't like John's definition because they claimed that:
(1) gravity varies depending on where you are on the earth and
(2) a change in gravity will affect the period of the pendulum.
    They were right about (1). The question is whether they were right about (2). Well . . . were they?

318. I'm pushing a giant pineapple across a frozen lake. If the strength of gravity changes, will that affect how hard I have to push?

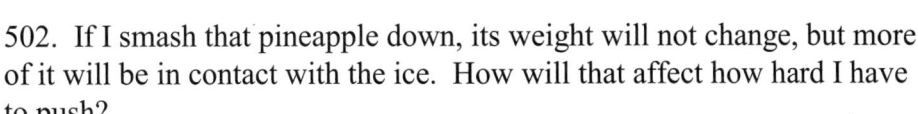

502. If I smash that pineapple down, its weight will not change, but more of it will be in contact with the ice. How will that affect how hard I have to push?

717. Is the number of bulldozers (🚜) I own a discrete or a continuous variable?

Chapter Four

c

First part: Problems from this chapter

119. In the metric system m stands for meter. Instead of writing 5 meters, you can write 5 m. 1 m ≈ 39 inches.* So a meter is a little longer than a yard. 1 m > 1 yard

We now have m and c. Which came first? Did we start with a meter and then define the speed of light, c, in terms of a meter, or did we start with the speed of light and define a meter, m, in terms of c?

455. When we are moving safes down the hallway, the formula is F = μW. F is the frictional force. It is how hard you have to push the safe to keep it moving at a constant speed. W is the weight of the safe. The constant of proportionality is μ. In the next chapter (Chapter 5) we will learn how to find μ.

Suppose you have this fellow who is just learning how to surf. He is standing on his board on the beach. He's nowhere near the water, but he's slightly frightened. You want to push him at 3 mph (miles per hour) along the sand so that he can feel what it's like to surf.

He and his board weigh 140 pounds. Mu (μ) between the board and the sand is 0.6. How hard will you have to push?

584. Suppose you want to push him at 9 mph instead of 3 mph. How hard would you have to push?

---

* Or better yet, 1 m ≈ 39.37 inches. But this is still just an approximation. The symbol ≈ means "approximately equal to."

Chapter Four c

Second part: the 𝔐ixed 𝔅ag: a variety of problems from this chapter and previous chapters

657. Our scared surfer falls over. He is clutching his surfboard. We tie a rope around him and his board and drag them. This will give him a taste of what it is like to fall off a board in the ocean.

After he was standing on his board, μ was 0.6. He is now lying in the sand holding his board. Will μ change?

805. The surfer had had enough "lessons" for the day. He put a rope on his surfboard and pulled it behind him. He was heading off to another part of the beach so that he could build sand castles.
Pulling the board behind him at a constant speed required 6 pounds of force.

Some kids tied a second board behind the first.

How much force did he need to apply now?

874. If the kids had piled the second board on top of the first,

Would that change how much force he needed to apply?

963. Light can travel one meter in $\frac{1}{299,792,458}$ of a second.

That's the same thing as saying it can travel 299,792,458 meters in one second. (It's fast.)

How far can it travel in 4 seconds?

Chapter Five
Finding Mu

First part: Problems from this chapter

160. Simplify $\dfrac{73\mu}{73}$

214. The formula for the coefficient of friction is $F = \mu N$.

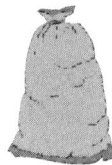

You have this large sack of carrots that you want to bring home to feed to your bunny. It weighs 240 pounds so you have to drag it rather than carry it.

Here is a picture of your situation.

sack on the sidewalk

You know that N is 240 pounds.

If * you knew the value of μ, you could find F, which the amount of force needed to drag that sack along the sidewalk.

For example, if μ were equal to 0.2, what would F equal?

395. But you don't know μ. There is no book or Internet site that will tell you the value of μ, the coefficient of friction between a sack of carrots and a sidewalk. Such a book would have to be a zillion pages long to list the coefficient of friction for every possible pair of surfaces in the world.

    Instead, you do the experiment. You drag your 240-pound sack of carrots and find that it takes 80 pounds of force to pull it at a constant speed. What is the value of μ?

---

* That's a big "if."

Chapter Five   Finding Mu

Second part: the 𝔐ixed 𝔅ag: a variety of problems from this chapter and previous chapters

492. You are given 36 = 5y. Find the value of y. (Hint: Divide both sides by 5.)

615. You are given 6x = 15. Find the value of x.

797. Given 30 = 7x and finding x is one of the first parts of algebra. When it is first taught, we put in all the baby steps . . .

Given	$30 = 7x$
Divide both sides by 7	$\dfrac{30}{7} = \dfrac{7x}{7}$
Simplify the right side	$\dfrac{30}{7} = x$
Simplify the left side	$4\dfrac{2}{7} = x$

After you have done 400 of these equations, you realize you can skip the second step. You go from 30 = 7x to $\dfrac{30}{7} = x$ (dividing both sides by 7).

Here are five equations. Do them slowly or blast your way through them.

20 = 5x   8y = 16   22w = 11   60 = 100v   4μ = 12

843. Your cute little bunny turns out to be a pig. He eats the whole sack of carrots. He now weighs 242 pounds.

It takes a hundred pounds of force to slide him across the kitchen floor. (You are heading toward the bathroom so that he can throw up.)
    What is the coefficient of friction, μ, of this fat rabbit and your kitchen floor?

17

Chapter Six
Super Fred

First part: Problems from this chapter

192. It takes 100 pounds of force to push your 242-pound rabbit across the kitchen floor. You weigh 120 pounds. What must the coefficient of friction, µ, be between your shoes and the kitchen floor in order for you to push that rabbit?

262.  It's New Year's Day and your mother made her traditional chicken soup for the family. It weighed 25 pounds. When you kid sister said, "Please pass the soup," you would push it across the table rather than lift it. If µ between the bottom of the bowl and the table was equal to 0.2, how much force would be needed to slide the bowl toward your kid sister at the rate of 6 inches per second?

744. Your kid sister removed the chicken from the soup. All that was left was the bowl and some warm water. Did this change µ?

902. Did this change the force necessary to slide the bowl?

924. Did this change your attitude toward your kid sister?

Chapter Six   Super Fred

Second part: the 𝔐ixed 𝔅ag: a variety of problems from this chapter and previous chapters

124. Find x.   100x = 37

211. This is an ellipse.

If you put a quarter  on the table and

look at it on a slant, it might look like this.

an ellipse!

Most adults don't know the area formula for an ellipse.

It is A = πab.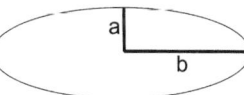

Suppose the area is 48 square inches and b is 8.
For this problem we will let π = 3.
How long is a?

486. Many adults know the area formula for a rectangle.

A stands for area.
ℓ stands for length.
w stands for width.

I use cursive ℓ for length because a printed l looks too much like a one.

If the area of a rectangle is 15 and the width is 10, what is its length?

19

Chapter Seven
Measuring Force

First part: Problems from this chapter

266. These are ordered pairs: (3, 7), (988, 16), (1.008, 62).
Why is this not an ordered pair? (8, 10, 16)

451. In 2 hours Fred can run 9 miles. Plot on a graph the point (2, 9).

595. In 1 hour Darlene can read 2 wedding novels.
In 2 hours she can read 4 wedding novels.
In 3 hours she can read 6 wedding novels.
In 4 hours she can read 7 wedding novels.
    Plot (1, 2), (2, 4), (3, 6), and (4, 7). Are these points on a straight line?

716. In graphing points you have a lot of freedom. If you want to graph the point (3, 14) you could draw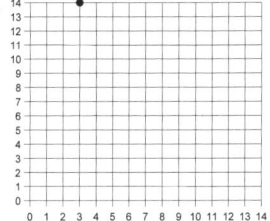

or you could stretch out the horizontal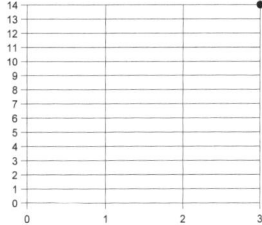

or you could simplify the numbering on the vertical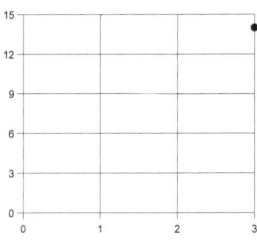

Chapter Seven   Measuring Force

or you could eliminate all those extra grid lines

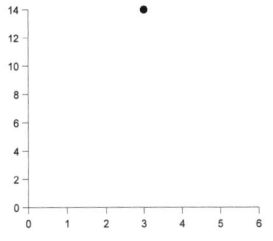

All of these are fine. Which one you choose is up to you.

You don't tell an artist which way to paint his pictures.

Plot the points (1, 3) and (20, 4).

872. What is the second coordinate of (20, 4)?

Chapter Seven   Measuring Force

Second part: the 𝔐ixed 𝔅ag: a variety of problems from this chapter and previous chapters

288. You have a chest of drawers. It is 5.1 feet tall. The area of the front is 14.28 square feet. What is the width?

5.1

457. It takes 10.8 pounds to slide your chest of drawers at 4 mph across the carpet in your bedroom. Your chest of drawers weighs 18 pounds. What is the coefficient of friction, μ?

555. If I slide your chest of drawers at 8 mph across the carpet in your bedroom, will the force needed be 5.4, 10.8, or 21.6 pounds?

601. Your kid sister removes one of the drawers from your chest of drawers.

She claims that she needed the drawer to hold her pet alligator. On her way to her bedroom she emptied the drawer in the hallway. "I want my 'gator to have lots of room to crawl around in the box," she said.

Your chest of drawers now weighs 13.5 pounds.

How much force will now be needed to slide it at 6 mph?

711. She drags her pet alligator into your bedroom and says, "I want you to apologize to my 'gator."
    "Why?" you ask. "You stole a drawer out of *my* chest of drawers. Why do I have to apologize to your pet?"
    "Because your drawer was too small for him."
    The alligator weighs 18 pounds. From all the information given on this page, can you find μ for the alligator and the carpet in your bedroom?

22

Chapter Eight
Absolute Truth

First part: Problems from this chapter

105. Back in Chapter 4 I wrote, "The Seventeenth General Conference on Weights and Measures rolled around in 1983. They were still talking about the length of a meter. They changed the definition again. **Since the speed of light in a vacuum is a constant**, they declared that a meter is the distance that light can travel in a vacuum in $\frac{1}{299{,}792{,}458}$ of a second."

Several centuries ago Isaac Newton wrote down what he thought was true. There was Newton's First Law, Newton's Second Law, etc. Since then, scientists did tons of experiments. No experiment overturned any of his laws. By 1900 virtually every scientist would have bet his two front teeth that Newton's laws were true.

In 1905 and 1915 Einstein showed that Newton's laws were *almost* true. He used the fact that the **since the speed of light in a vacuum is a constant** and deduced surprising things such as when things are moving near the speed of light they (1) get more massive and (2) time slows down for those things.

Since Einstein's work virtually all scientists would be willing to bet their eyebrows that **the speed of light in a vacuum is a constant**.

It's a law of physics.

Make a guess. What are physicists discussing nowadays?

218. List three things that scientists have proven to be 100% certain.

420. Your kid sister drags her pet alligator into your bedroom. You can see it. You can hear it when it snaps its jaws. You can lick it and taste that it is an alligator. You can smell it. (Your sister forgot to wash it.) You can feel it when it bites your arm.

continued next page

Chapter Eight   Absolute Truth

Your five senses—sight, hearing, taste, smell, and touch—are *the only ways you know* what is out there in the world beyond yourself.

These five senses are the only ones that scientists have.

All five of your senses tell you that it's an alligator.

Can you be 100% certain?

570. A hunch might only have a 5% chance of being true.

A conjecture might only have a 30% chance of being true.

A theory might only have an 80% chance of being true.

A law might only have a 98% chance of being true.

You wake up in the morning. You are in your bed. You decide that you are only going to do things where you are 100% certain of their outcome.

    i) Do you get out of bed and walk across the room?

    ii) Do you pour some cereal into your bowl?

    iii) Do you ask Pat* to marry you?

---

* Pat is either Patrick or Patricia.

Chapter Eight   Absolute Truth

Second part: the 𝔐ixed 𝔅ag: a variety of problems from this chapter and previous chapters

315.  Your kid sister has attached her pet 'gator to the ceiling with a spring.
The formula is $F = kx$, where F is the force on the spring (in this case, the weight of the alligator), where x is the distance the spring has stretched, and where k is a constant.

Suppose the alligator weighs 18 pounds and the spring has stretched 20 inches. What is the value of k?

498. Now suppose the alligator eats your kid sister's sticker collection. He now weighs 27 pounds. How far will the spring stretch?

643. Your kid sister's sticker collection was world-famous. Before her alligator ate all her stickers, she owned 8,398,200,335 stickers. Now she owns 0 stickers. Is the number of stickers owned a discrete or a continuous variable?

763. When that alligator is not tied up or hanging from the ceiling, your sister lets it run around in her bedroom. Sometimes it is 7 feet from her bedroom door. Sometimes it is 1 foot from her bedroom door. Is the distance that the alligator is from her bedroom door a discrete or a continuous variable?

903. You have just bought a lovely bit of property in Kansas. It has a house with some trees around it and miles of open land for you to ride your horse on. The real estate agent told you that you own 18.6 square miles in the shape of an ellipse.

The area formula for an ellipse is $A = \pi ab$.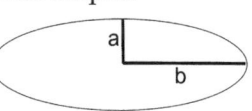
The distance from the center of your land to the north end is 2 miles. (That means that a is 2.)
How far is it from the center to the east end? (Translation: How long is b?)
For this problem we will let $\pi = 3.1$.

25

Chapter Nine
Four Ways to Stretch

First part: Problems from this chapter

> When you have an alligator on a spring and it doesn't weigh too much, then you are in the region that goes up to the proportional limit.
> With more weight you are in the region that goes up to the elastic limit.
> With even more weight you enter the plastic region. Finally, with a super-heavy alligator you reach the breaking point.

189. How much can you stretch that spring so that after you remove the alligator the spring will be good as new?

352. In what region or regions will Hooke's law hold?

536. Let's suppose that you have a really large number for the spring constant. You have a spring where F = 1,000,000x where F is measured in pounds and x is measured in inches.

　　If you hang a ten-pound weight on this spring, how much will the spring be stretched?

673. Your sister attached her alligator to the ceiling with a big spring. "Nothing but the best for my 'gator," she shouted.

　　The alligator had eaten your sister's doll collection and now weighed 60 pounds. It stretched the heavy spring 6 feet. The spring constant, k, is 10. The math: F = kx ⇨ 60 = k6 ⇨ 10 = k

　　If it had also eaten your sister's furniture, it would weigh 200 pounds. The math says that the spring would stretch 20 feet.

　　The heavy spring has a proportional limit of 400 pounds, so we would expect Hooke's law to hold.

　　The math: F = kx ⇨ 200 = 10x ⇨ 20 = x.

　　But there was something wrong. The spring only stretched 9 feet. If you were standing there, you would know immediately what went wrong. Why did the spring stretch only 9 feet?

26

Chapter Nine   Four Ways to Stretch

Second part: the 𝔐ixed 𝔅ag: a variety of problems from this chapter and previous chapters

369. Simplify $\frac{893z}{893}$

490. You showed your kid sister a picture of the property you bought in Kansas. It is in the shape of an ellipse. You explained to her that the area of an ellipse is A = πab.

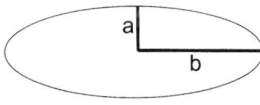

"Ha! Ha! Ha!" your sister shouted. "You can't eat land." I just made a plate of food for my 'gator. He will love it. And my plate is round like a circle. It's not like a stupid ellipse—a squished circle. Ellipses aren't as much fun as circles."

You point out that circles are just ellipses in which the a and b distances are the same. If you know that $A_{ellipse}$ = πab, then what must be the area of a circle?   This is an easy question.

816. You kid sister didn't like thinking of her circular plate as just being a special case of ellipses.
  She shouted, "You see that black line around the edge of my plate?" (She didn't know that the distance around the edge of a circle is called its **circumference**. That's okay. She's only 3 years old.)
  The formula for the circumference of a circle is C = 2πr.
  Given an ellipse, can you give the formula for the distance around the edge (the perimeter)?

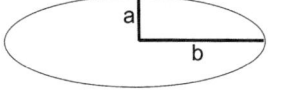

Warning: This is an impossibly hard question for anyone who hasn't had at least a year of calculus.

27

Chapter Ten
Two Kinds of Friction

First part: Problems from this chapter

122. Your kid sister came into your bedroom and pulled out all the drawers in your chest of drawers.

This time she didn't empty out the drawers in the hallway. She just stacked them on your bed.

"I need a big empty box," she said. She didn't ask your permission. "I need it for my 'gator."

Your empty chest of drawers now weighed 9 pounds. It would take 6 pounds of force to get it moving. Find μ_s.

317. You don't want your sister to take your chest of drawers. You find your heaviest clothes, which is a suit of armor, put it on, and stand on top of the chest.

Now the whole thing weighs 171 pounds.

Will μ_s have changed?

406. How hard will your sister have to push to get you and the chest moving?

572. From the previous problem we know how hard your kid sister will have to push to get you and the chest moving. She's wearing her princess gown. (These were the outfits you wore on Halloween.) μ_s between her shoes and the carpet in your bedroom is 0.3. She's putting rocks in the pockets of her gown. How much will she and her gown have to weigh to move you?

Chapter Ten   Two Kings of Friction

Second part: the 𝔐ixed 𝔅ag: a variety of problems from this chapter and previous chapters

655. It would be crazy for your sister to get enough rocks in her pocket so that she weighed 380 pounds.

   Her three-year-old mind was hard at work in order to get ~~your~~ her box out of your bedroom and into hers.

   Several thoughts occurred to her . . .

❖ Call Mom and have her make you get off the chest. That wouldn't work since she would say that it really is your chest. She would call me naughty.

❖ Throw a pillow at you to knock you off. That wouldn't work. The pillow would just bounce off his armor.

❖ Quit stealing your chest. Ha! I'll never do that. It was your chest. It's now mine.

   She could think of only one solution. She headed into the backyard and dragged the garden hose down the hallway and into your room. You wouldn't get off, so she let you have it.

   In the hallway the hose weighed 4.8 pounds and μ_k was 0.7. How much did she have to pull on the hose to keep it moving?

842. In one hour she might steal two things from you. (1, 2). In 3 hours she might take 6 things. (3, 6). Graph these two points. Draw a line through them. Estimate how many things she'll steal in 4 hours.

981. She made a giant puddle of water on your bedroom floor. It was in the shape of an ellipse. It measured 6 feet across at its widest and 4 feet at its narrowest. What was the area of this puddle? Use $\pi = 3$ for this problem.
(Hint: The answer is not 24π or 72.)

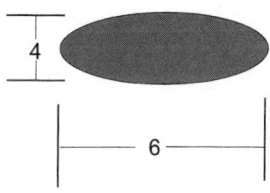

Chapter Eleven
Energy

First part: Problems from this chapter

158. Your kid sister has a big rubber ball. It is so big that her alligator can't eat it. At night she likes to bounce her ball in the hallway.

When it's heading downward toward the floor, it has what form of energy?

367. When it hits the ground, part of the energy is turned into sound, which keeps everyone in the house awake.

But at the very instant it hits the ground and squishes a bit, it is neither traveling downward nor upward. There is no energy of motion at that point in time. Besides the energy of sound, what has the motion energy been turned into?

428. The sound of that big rubber ball bouncing (bounce, bounce, bounce, bounce, bounce, bounce, bounce, bounce, bounce, bounce, bounce, bounce, bounce, bounce) hits your ears and is sent to your brain by the nerves that connect your ears to your brain.
Your brain thinks. Where does it get its energy from?

Chapter Eleven   Energy

Second part: the 𝔐ixed 𝔅ag: a variety of problems from this chapter and previous chapters

625. Your three-year-old sister gets a big thrill from dropping her big rubber ball and watching it
bounce, bounce, bounce, bounce, bounce until it finally lies quietly on the floor.

    Each time the ball hits the hallway floor, the motion of energy is turned into spring energy, sound energy, and heat energy. (The ball gets slightly warmer as it bounces.)
    Suppose she drops her ball from 8 feet. Because of the loss to sound and heat, it bounces back to half its height, 4 feet.
    $8 \times \frac{1}{2} = 4$
    Then it bounces back to 2 feet. $4 \times \frac{1}{2} = 2$
Do the arithmetic. When will it be bouncing less than $\frac{1}{100}$ of a foot?

722. You are in your bedroom with the door shut. It's 11 p.m. You hear the familiar bounce, bounce, bounce, bounce, bounce, bounce, bounce, bounce, bounce, bounce, bounce, bounce, bounce, bounce, bounce. For the last week your kid sister has been out in the hallway bouncing her big rubber ball in the middle of the night. Every time you have gotten out of bed and gone into the hallway, she has apologized and said that she'll never do it again.
    She's three years old and forgets her promises.
    You feel anger. You feel frustration. You have complained to your parents, and they have done nothing. You think about 23 years from now when she'll get married and move out of the house, so that you'll have a decent night's sleep.
    How certain are you that she's out there in the hallway bouncing her ball?

817. Find x. $50 = 80x$

919. You attach that 18-pound alligator to the ceiling with a spring. It stretches the spring 1.2 meters. (A meter is a little longer than a yard.) You then take the big rubber ball, which weighs 6 pounds) and balance it on the alligator's nose. How long will the spring be stretched now?

31

Chapter Twelve
A Time for Action

First part: Problems from this chapter

126. In the book Fred had passed out and was on top of the broken door that lay on the floor. Kingie had to push with a force of 21 pounds to get the door moving with Fred on it.
    The coefficient of static friction, μ_s, between Kingie's shoes and the floor is 0.7.
    Kingie is too light to push Fred and the door. He only weighs one pound. How many pounds of art supplies must be put in his pocket to get Fred and the door moving?

261. To fix the broken door, the carpenter submitted this estimate:

Labor	$1,800.00
Materials	$ 186.92
7% sales tax on materials	?

How much was the sales tax?

501. (continuing the previous problem) What is the total bill from the carpenter?

751. Kingie volunteered to pay for the damage to the office door. His paintings often sell for $8,000. What *percent* of the sale of one of his paintings will cover the cost of repair?
    (Translation: $2,000 is what percent of $8,000?)

876. If it takes Kingie 57 minutes to do an oil painting (such as "Swiss Mouse"), how long will it take him to do six of these paintings?

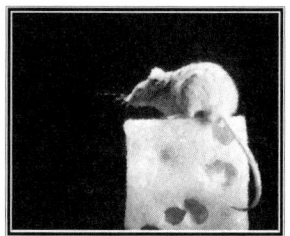

"Swiss Mouse"
by Kingie

Chapter Twelve   A Time for Action

Second part: the 𝔐ixed 𝔅ag: a variety of problems from this chapter and previous chapters

194. The alligator couldn't eat the big rubber ball, but it did bite it. Now it is just a six-pound glob of rubber on the floor.

You want to drag that glob off to the garbage can. If it takes 4.2 pounds to get it moving, what is the coefficient of static friction between that glob and the floor?

263. If the alligator hops on top of the glob, will μ_s increase, decrease, or stay the same? The alligator weighs 18 pounds.

360. How much force will it take to get the glob and alligator moving?

453. It takes 10 pounds to move the glob and alligator at 8 mph across the floor. How much force is needed to move them at 16 mph?

750. Your kid sister really doesn't like your hauling her pet alligator and her big rubber ball off to the garbage can. Instead of squirting you with a hose, she sits on the glob and cries like a three-year-old. Glob, 6 pounds. Alligator, 18 pounds. Sister, 24 pounds. How much force will be needed to pull the glob + alligator + sister at a constant speed toward the garbage can? (Hint: You'll need to use information from the previous problem.)

953. For the last three years every time you do something that your kid sister doesn't like, she cries. You sing too loudly, she cries. You don't apologize to her alligator, she cries. You smile, she cries. You don't read her favorite book to her ('*Gators Are Cute*), she cries. You complain to Mom when she steals a drawer from you chest of drawers, she cries.
   You figure that when you get to the garbage can, she'll be crying when you throw the glob away. Is this an example of inductive or deductive reasoning?

Chapter Thirteen
Flabbergasted Fred

First part: Problems from this chapter

319. You didn't throw the alligator into the garbage can. You are too nice to do that to your kid sister's pet.
   The next morning when you went into the kitchen to have your breakfast, there it sat on the kitchen table.
   You asked it to get off, but alligators are very poor at understanding English.
   You knew that you just couldn't ignore it. The minute you sat down with your breakfast, it would probably eat it.

You decide to tip the table and make it slide off.

Here is the moment in which
it began to slide.

Compute the coefficient of static friction, μ_s, between the alligator and the kitchen table top.

❀ Express your answer as a fraction.
❀ Express your answer as a decimal rounded to the nearest one-hundredth.

34

Chapter Thirteen   Flabbergasted Fred

Second part: the 𝔐ixed 𝔅ag: a variety of problems from this chapter and previous chapters

594. With the alligator off the kitchen table you could sit down and enjoy a box of Wizard shredded cereal.

Time out. You will learn more about Wizard shredded cereal in Chapter 21 of Life of Fred: Calculus Expanded Edition in which we learn about partial derivatives. In that chapter the expanded chain rule with intermediate variables is $\frac{\partial z}{\partial t} = \frac{\partial z}{\partial x}\frac{\partial x}{\partial t} + \frac{\partial z}{\partial y}\frac{\partial y}{\partial t}$

Of course, none of that makes any sense at all right now, but once you finish the three Life of Fred pre-algebra books,
        beginning and advanced algebra,
      geometry,
        and trig,
you can start college calculus. Then in the 21ˢᵗ chapter Fred and his friends—the scarecrow, the lion, and the tin-wood duck—will be visiting Wizzie, the Wizard of Real Estate, and enjoying his cereal. At that point, all that fancy chain rule stuff will be as clear as μ_s *= rise/run is to you today.*

    Today it takes three drops of sweat to learn about static friction. When you get to calculus, it will still take three drops of sweat to learn the new stuff.

    When you kid sister came into the kitchen,* she noticed that her alligator was on the floor. She, of course, cried and put it back on the kitchen table. Before it could eat your box of Wizard shredded cereal, you pulled the box toward you.** It took one ounce of force to pull the three-

---

* A common English error is to say, ". . . came in the kitchen."

** In American English, *toward* is more common. In British English, *towards* is more frequently encountered. English is hard enough (it's harder than math), but when you deal with American vs. British English, things get even harder. Americans write *gray*, and the British write *grey*.

Chapter Thirteen   Flabbergasted Fred

ounce box toward you at the constant speed of 1 inch per second. Do you have enough information to compute either μ_s or μ_k?

723. You grab your box of Wizard shredded cereal and head into the living room. The alligator jumps off the table and follows you. Your sister screams, "You're taking my alligator!"

Mom arrives. Now you know that things will get better. Everything will be fair. You are wrong.

Mom shouts at you, "What are you doing eating in the living room? I've told you a thousand times to eat in the kitchen."

You head into the kitchen and put your box of Wizard shredded cereal into the cupboard,* and head off to your bedroom. You shut the door.

You were tired of eating in the same room with your kid sister (and her alligator). You did the logical thing: you installed a kitchen in part of your bedroom. It has all the things that a regular kitchen has. In addition, it has a giant pizza oven. (See where the black arrow is pointing.)

You have stocked your bedroom kitchen with cases of Wizard shredded cereal, gallons of milk, and many frozen pizzas. If you want to, you could stay alone in your bedroom for weeks.

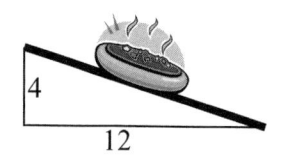

You invent the Pizza Ramp. You put your hot pizza on the slide and gradually increase the rise until the pizza starts to move.

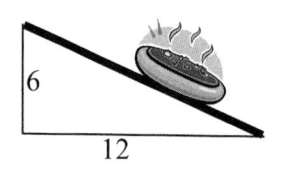

When the rise was equal to 6, it moves.

---

* The *p* in *cupboard* is silent just like the *k* in *knife* is silent. None of the symbols in 2 + 2 = 4 are silent.

Chapter Thirteen   Flabbergasted Fred

The pizza slides down onto the table. It stops right in front of you and is ready to be eaten.

The pizza weighs 4½ pounds. You take your fork and push it an inch to the left. How much force will you need to apply to get it moving?

904. The pizza is made out of rubber. Your kid sister switched the pizzas you ordered for a Jolly Joker Joking Pizza. When you bite down on it, it acts like a spring.

When you bite with 10 pounds of force, your teeth sink 0.4 inches into the pizza. If you bite with 6 pounds of force, how far will your teeth sink into it?

Ha! Ha! Ha! Ha! Ha!
Ha! Ha! Ha! Ha! Ha!
Ha! Ha! Ha! Ha! Ha!
Ha! Ha! Ha! Ha! Ha!
Ha! Ha! Ha! Ha! Ha!

Your sister's laughter

Chapter Fourteen
Normal Force

It takes three chapters in *Life of Fred: Pre-Algebra 0 with Physics* to show how to compute the coefficient of static friction, μ_s, without knowing the weight of the object—without $F = \mu_s N$.

✸ In Chapter 13 Kingie declares μ_s is equal to $\frac{\text{rise}}{\text{run}}$

✸ In the current chapter, Chapter 14, we learn a bit of art. We learn how to resolve a weight W on a tilted surface into N and F.

✸ In the next chapter, Chapter 15, we show why $\frac{F}{N}$ equals $\frac{\text{rise}}{\text{run}}$

Or, more slowly, we start out by knowing that μ_s is equal to $\frac{F}{N}$ because we know that $F = \mu_s N$.
Then in Chapter 15 we show that $\frac{F}{N}$ equals $\frac{\text{rise}}{\text{run}}$
These two things together show that μ_s is equal to $\frac{\text{rise}}{\text{run}}$ which is what Kingie claimed in Chapter 13.

First part: Problems from this chapter

285. Draw each of these on a sheet of paper. (Do not write in this book.) Then sketch in N and F.

You can either trace these pictures or just draw them on your sheet of paper.

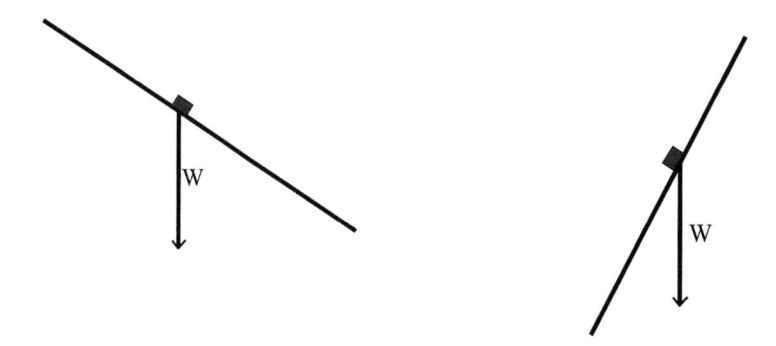

38

Chapter Fourteen   Normal Force

Second part: the 𝔐ixed 𝔅ag: a variety of problems from this chapter and previous chapters

456. Which is the larger number, 3 or 9?

544. In the metric system 6 m stands for six _____.
fill in one word

641. You are given 8x = 7. Find the value of x. Express your answer both as a fraction and as a decimal.

760. The front door to your bedroom is three feet wide and six feet tall. Your kid sister knew that you would like it if she decorated your bedroom door with a picture she drew. She drew the biggest ellipse that would fit on your door and painted it in.

What is the area of that ellipse?
(Use 3 for π in this problem.)

841. She painted right over the sign you had put on your door.

The sign was 2.9 feet wide and 1.6 feet tall. What is the area of your sign?

39

Chapter Fifteen
Math Makes Things Easier

First part: Problems from this chapter

We will use these facts from geometry:
- ✴ The angles of any triangle always add up to 180º
- ✴ Every right angle is equal to 90º
- ✴ In geometry angle 3 is often written as ∠3.
- ✴ Triangle ABC is often written as △ABC

Suppose we are given this diagram:

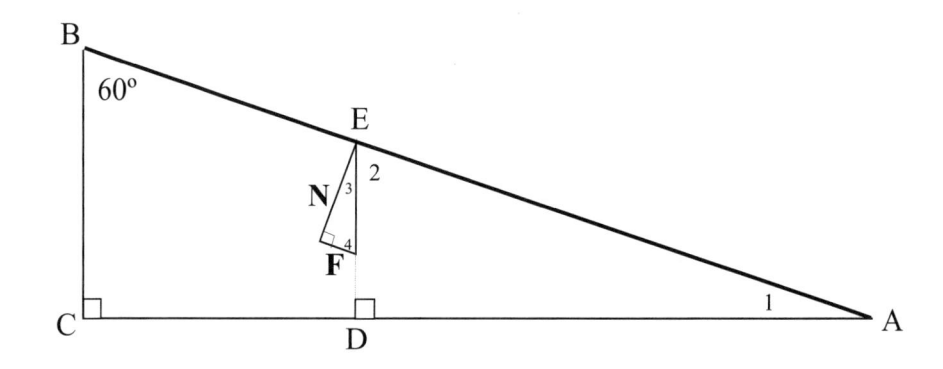

200. How many degrees is ∠1?

596. How many degrees is ∠2?

660. How many degrees is ∠3? (Note that the vector **N** is at right angles to segment BEA.

720. How many degrees is ∠4?

40

Chapter Fifteen   Math Makes Things Easier

Second part: the 𝔐ixed 𝔅ag: a variety of problems from this chapter and previous chapters

163. You talk to your dad. "Did you see what my kid sister did? She painted the door to my bedroom!" You know that now things will get better. Everything will be fair. You are wrong.

You and he walk to your bedroom door. He says, "That's not bad work for a three-year-old."

"But Dad! That's *my* door."

He scratches his head and suggests, "Well, to be fair, you should get to paint her bedroom door."

"But Dad! I don't want to paint *her* door. I just want to have what is mine."

You head into your room and shut the door. You don't slam it, even though you are angry. Your kid sister slams doors all the time.

You tape up an alligator target to the wall in your bedroom.

It has a diameter of 3 feet. What is the radius of that circle?

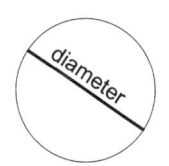

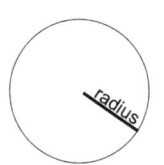

260. You throw boogers at that target. What is the area of that target?

383. You take the target off of the wall and drag it toward the garbage can. It weighs 8 pounds and takes 3 pounds of force pull it across your bedroom floor. Find μ_k.

489. As you drag it across the floor, some of the boogers that are on top of the target (not in contact with the floor) drop off.

You are now pulling a load of 7 pounds instead of 8. How many pounds of force will it take to get it moving at 2 mph?

41

Chapter Sixteen
Work

First part: Problems from this chapter

267. *Ounces* in the Imperial system (the one used in the United States) can mean either a weight or a volume. The metric system (the one used in the rest of the world) doesn't have this confusion. A gram (weight) is a gram, and a liter (volume) is a liter.

Now the fun begins. If you buy 5 ounces of chocolate and you buy 5 ounces of silver, which weighs more?

(Hint: Almost everything is measured in regular ounces.* One ounce = 28.349527 grams. However, silver (and gold) are measured in troy ounces. One troy ounce = 31.103481 grams.)

714. How much more? (Use one avoirdupois ounce ≐ 28.3 grams and one troy ounce ≐ 31.1 grams.)   ≐ means "equals after rounding off"

845. You lift that 5-ounce bar of chocolate from the table up to your nose. (That's an 18-inch trip.) There are few things that smell that good. How many ft-lbs of work did you do?
(1 foot = 12 inches; 1 pound = 16 ounces)

918. Recall the Giant Chart of Energy back in Chapter 11. Physicists tie those nine forms of energy to their concept of work. The formula Work = force times distance involves changing one form of energy into another.

Suppose for the sake of discussion you put a stick of dynamite under somebody's alligator and blew it 50 feet into the air. Since it weighs 18 pounds, you will have done 900 ft-lbs of work. What form of energy did you change into what other form?

Of course, you would never do that if it were your kid sister's pet.

---

* If you want to impress your friends, the official name for regular ounces is avoirdupois. Pronounced ever-deh-POIZ.

Chapter Sixteen   Work

Second part: the 𝔐ixed 𝔅ag: a variety of problems from this chapter and previous chapters

123. You lift that 7-pound target three feet upward (not upwards) and put it into the garbage can. How much work was done? (a one-step problem)

312. You pull the garbage can down the hallway at 3 mph. It weighs 20 pounds and takes 8 pounds of force to pull it.
   As you pass your kid sister's door, she opens the door and drops a 40 pound package of used Halloween candy into the garbage can.
   How much work will you now do in dragging the can 15 feet down the hallway?
   This question is a three-step problem.

497. You lift that 5-ounce bar of silver 18 inches upward so that you can get a good look at it. Does that involve the same amount of work as when you lifted your 5-ounce bar of chocolate 18 inches upward to take a sniff of it?

799. You put your silver in your safe, which is right next to the kitchen in your bedroom.
   You hang your 5 ounces of chocolate on a 4-foot string attached to the ceiling and start it swinging. You measure the period (the time it takes to make a complete trip back and forth).

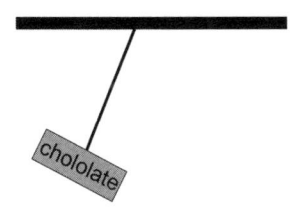

As it swings over near you, you take a bite out of the bar. Does this increase or decrease the period?

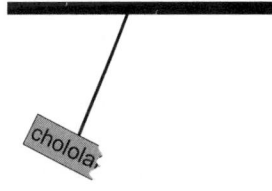

925. You put the half-eaten bar of chocolate into your microwave oven and set it for 40 seconds. It is turned into a warm, gooey, delicious puddle. Energy has been converted from what form into what form?

Chapter Seventeen
Transfer of Energy

First part: Problems from this chapter

156. Your kid sister is in her bedroom with the door shut. She is watching her kiddie programs on television.
   Sammy Spider Sings Songs
   Mary Makes Muddy Pies
   Tough Tim Toots a Tune
   Harry Hums

   The noise that comes out of the room is driving you nuts. She has these programs going all the time.
   Which of these "facts" is not true?
Fact A: Energy in the form of sound is leaving her bedroom.
Fact B: The Law of Conservation of Energy: In any closed system, the amount of energy cannot change.
Fact C: Since energy (sound) is coming out of her bedroom it is not a closed system.
Fact D: If you have a system in which energy is only leaving the system, then the amount of energy in the system must decrease as time goes on.
Fact E: There is no energy entering her bedroom.

289. You knock on her door and ask her to turn her television down. She responds, "I can't hear you. Talk louder."
   You yell, "Please turn your TV down a bit. I'm trying to read."
   Your mom comes. "Why are you yelling? Please use your inside voice when you are in the house."
   You are about to make the very logical argument . . .
   She asked me to speak louder because she couldn't hear me. She couldn't hear me because she's got those silly kiddie programs turned up so high. When I'm in my bedroom minding my own business, I am entitled to the peaceful enjoyment of my space. In law, her excessive noise is called a trespass. No one's freedom extends to the point of decreasing other people's freedom.
   You didn't get a chance to say this. Instead, your mom pulled you down the hallway and into your room. The work of her pull (from the chemical energy stored in her body into motion) was then transformed into what other form of energy?

Chapter Seventeen   Transfer of Energy

Second part: the 𝔐ixed 𝔅ag: **a variety of problems from this chapter and previous chapters**

287. It was fun to play in your kitchen in your bedroom. You timed yourself. You walked into your kitchen, and in 5 minutes you could make 6 waffles. On another day you found that you could make 8 waffles in 6 minutes. Plot these two points (5, 6) and (6, 8).

534. Looking at the graph you drew, estimate how many waffles you could make in 8 minutes.

712. The speed of light in a vacuum is exactly 299,792,458 meters per second. In this Chapter 17 we mentioned Einstein's famous equation $E = mc^2$. E is energy, m is mass (or matter), and c is the speed of light.

$E = mc^2$ means that a little bit of matter is equal to a huge amount of energy. Suppose m were equal to, say, 5. What would E be equal to?

For this problem, to make the arithmetic a little easier, suppose that c is equal to 300,000,000 rather than 299,792,458.

873. Kingie made $\frac{rise}{run}$ famous. If you tilt a flat surface upward until the block begins to slide, then the coefficient of static friction, μ_s, is equal to $\frac{rise}{run}$

Let's go 𝕆𝕟𝕖 𝕊𝕥𝕖𝕡 𝔻𝕖𝕖𝕡𝕖𝕣.

We will head into the land of trig (also known as trigonometry). Suppose you know that ∠A is equal to 23°.

After
  Life of Fred: Pre-Algebra 0 with Physics,
  Life of Fred: Pre-Algebra 1 with Biology
  Life of Fred: Pre-Algebra 2 with Economics
  Life of Fred: Beginning Algebra Expanded Edition
  Life of Fred: Advanced Algebra Expanded Edition
  Life of Fred: Geometry Expanded Edition

will come *Life of Fred: Trig Expanded Edition*. It is the course right before college calculus.

In the land of trig, we will learn about the tangent function, which everyone abbreviates as tan. (It has nothing to do with getting a sunburn.)

45

Chapter Seventeen Transfer of Energy

In trig, we will define tan A to equal $\dfrac{\text{rise}}{\text{run}}$

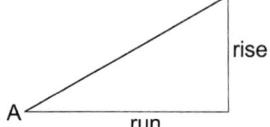

So if ∠A is equal to 23°, then tan 23° = __?__ in terms of friction.
This is not a super easy question. You may have to read the question several times before the answer becomes clear.

The answer is one step beyond $\dfrac{\text{rise}}{\text{run}}$

990. To solve 20 = 5x you divide both sides by 5 and get 4 = x.

Now for something new. To solve 13 = x + 6 you subtract 6 from both sides and get 7 = x. Neat, isn't it?

You do five of these. By the time you get to the fifth one, you should be doing these as fast as you can write.

Solve   18 = x + 2

200 = y + 50

w + 4 = 11

6 + z = 60

33 = 30 + x

Chapter Eighteen
Storing Energy

First part: Problems from this chapter

106. This is a true sentence: One thousand calories is equal to one Calorie.
If you read that sentence aloud, you will sound really stupid. Anyone listening to you will ask, "How can one thousand of something equal one of that same thing?"

A calorie is sometimes called a **small calorie**.

A Calorie is different than a calorie. A Calorie is used to measure the energy in food. A Calorie is a thousand times bigger than a calorie.

(Have I ever mentioned that English is harder than math? When you get to beginning algebra, the word problems are often the hardest part of algebra. You have to turn English into equations. Once you get the equation, then doing the algebra is duck soup.* Once you get to
x + 10 = 27, then it's one step to x = 17—you just subtract 10 from both sides.)

In any physics handbook, you'll probably find all kinds of conversions. One of the popular ones is 1 Calorie = 3,088 ft-lbs.

Whenever we want to convert from Calories to ft-lbs or from ft-lbs to Calories, we will use one of these **conversion factors**:

$$\frac{3088 \text{ ft-lb}}{1 \text{ Calorie}} \quad \text{or} \quad \frac{1 \text{ Calorie}}{3088 \text{ ft-lb}}$$

Both of these fractions are equal to 1 since the numerator and denominators are equal.

So, multiplying by either of these won't change the value, since multiplying by 1 never hurt anything.

Suppose I start with 2777 ft-lb of work and want to change that into Calories. I choose the second conversion factor so that the units will cancel. (continued on next page)

---

* duck soup = easy, no sweat, a piece of cake

Chapter Eighteen   Storing Energy

$$\frac{2777 \; \cancel{\text{ft-lb}}}{1} \times \frac{1 \; \text{Calorie}}{3088 \; \cancel{\text{ft-lb}}}$$

and what I have left is Calorie, which is what I wanted.

Doing the arithmetic $= \dfrac{2777}{3088}$ Calories

Your question: Change 7 faradays into ampere-hours using a conversion factor. (In physics handbooks, you can learn that 26.8 faradays equals one ampere-hour.)

364. In the old days before fast food restaurants, dying of starvation was much more common. Summertime often had more food available than winter.

a fruit tree I planted years ago   winter at my house

     In order to survive the winter we needed to store energy in our bodies. One gram of fat is equal to 9 Calories. One gram of protein is equal to 4 Calories. We can store energy (Calories) in our muscles, but fat storage makes much more sense.

     Three pounds of protein (muscle) can store as much Calories as how many pounds of fat? (1 pound ≈ 453 grams)

Chapter Eighteen   Storing Energy

Second part: the 𝔐ixed 𝔅ag: a variety of problems from this chapter and previous chapters

161. Solve 8.9 = x + 3.94

392. Solve 7y = 48.3

564. The pizza oven in the kitchen in your bedroom is a genuine wood-fired oven imported from Freedonia. Burning oak gives a much different flavor than using either gas or electricity. There is energy stored in the oak that is released as heat when it burns. Which of the nine forms of energy is used to store energy in the oak?
motion/heat/light/sound/electrical/height/nuclear/spring/chemical

709. You want a pizza snack. You roll out the dough. Since you are new to making pizzas, it turns into an ellipse rather than a circle. It's 2 feet by 1 foot. To the nearest tenth of a square foot, what is its area?

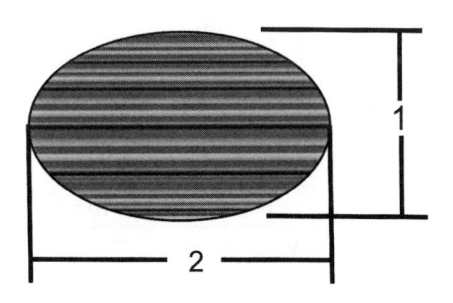

718. Convert 1.6 square feet into square inches. Use a conversion factor. 144 square inches = 1 square foot.

974. A 16" pizza is a large pizza in most pizza restaurants. The 16" is the measure of the diameter. What is the area of that pizza?

Chapter Nineteen
Metric System

First part: Problems from this chapter

157. Only the U.S.A. and one tiny country in Africa still use the Imperial system with 2 cups = 1 pint; 2 pints = 1 quart; 4 quarts = 1 gallon; 32 (dry) quarts = 1 bushel.

The metric system is so much cleaner with 10 millimeters = 1 centimeter; 100 centimeters = 1 meter; 1000 meters = 1 kilometer. To do conversions you often just move the decimal point.

In this chapter you were warned that good old pounds in the Imperial system as a measure of force doesn't correspond to kilograms. Kilograms are a measure of mass. In the metric system force can be measured in newtons. This was named after Isaac Newton, not fig newtons.

Time in the Imperial system is measured in seconds. Have you ever been told what the unit used for time is in the metric system?

491. Joules (pronounced JEWELS) is a measure of work in the metric system. What is the measure of work in the Imperial system?

639. On Mars it is true that 1000 meters = 1 kilometer.
On earth 1 kilogram ≈ 2.205 pounds. Is that true on Mars?

762. Let's try to get everything straight.
    Work----------ft-lb (in Imperial)----------joules (in metric)
Force or weight----------pounds (in Imperial)----------newtons (in metric)

Any physics handbook will tell you that 1 pound ≈ 4.448 newtons.
Using a conversion factor, translate a push of 5 pounds into the metric system.

Chapter Nineteen   Metric System

Second part: the 𝔐ixed 𝔅ag: a variety of problems from this chapter and previous chapters

209. Your elliptical pizza is nicely cooked in your wood-fired oven. You stab it with a BBQ fork and drag it out of the oven (at a constant speed). If you know the pizza weighs 6 pounds, what else would you need to know in order to determine the constant of kinetic friction, μ_k?

449. Suppose you could tip the oven rack and found that the pizza would start to slide when the rack was tipped at 27°. Is that enough information for you to determine μ_k?

656. You kid sister knocks on your bedroom door and says, "I smelt your pizza.* Can I have a piece?" You give her a 1.7-pound slice of your 6-pound pizza. How much is left for you?

798. 1.7 is what percent of 6? Round your answer to the nearest percent.

819. Solve $y + 8.88 = 9.99$

929. Now you are ready for a two-step algebra equation.
We want to solve $3x + 5 = 32$.
The first step is to subtract 5 from both sides.

---

* Both *smelt* and *smelled* are correct past tenses of the verb *smell*. If you have a decent dictionary, you can look up *smell*, and your dictionary will tell you the two past tenses of that word.

Chapter Twenty
Measuring Mass

First part: Problems from this chapter

212. We know that on earth 1 kg weighs (approximately) 2.2 lbs. For your pizzas you need a lot of pepperoni. You order 18 kg of pepperoni from Uruguay.

The climate of Uruguay is like southern California. There aren't many countries with three u's in their name.

How many pounds of pepperoni have you ordered? Use a conversion factor.

To remember where Uruguay is in South America, think Under.

485. You are talking to your friend on Mars who has never been to earth. You are trying to explain over the telephone exactly how much a kilogram is. The only way you can communicate is by voice. You can't mail him a 1-kilogram bar.

Why wouldn't this approach work? *Find something that weighs the same as 2.205 pounds. That's a kilogram.*

Intermission

So there are parts of physics that couldn't be communicated just by voice.

I'm going to claim that every part of math could be taught over the phone. It might be hard, but it could be done. If we were to receive radio messages from some super smart civilization in another galaxy, we could chat about solving 6x + 11 = 13 or knowing that an implication and its contrapositive are logically equivalent. **Stop! I, your reader need to**

52

Chapter Twenty   Measuring Mass

interrupt you. I have a question and an objection.

Okay. Let me hear them.

First, my question. Actually, three questions.

Q1: What is an implication?
Q2: What is a contrapositive?
Q3: What does logically equivalent mean?

An implication is an if–then statement. *If it is a raven, then it is black.* R ⇒ B.

The contrapositive of *If it is a raven, then it is black* is *If it is not black, then it is not a raven.* not-B ⇒ not-R.

These two statements are logically equivalent means that either statement is true if and only if the other statement is true. If you know either statement is true, then the other statement must be true.

We cover all this stuff in *Life of Fred: Logic.*

You said you have an objection.

My objection is simple. You could talk to your monster friends on some distance star, but you couldn't do geometry. Ha! How are you going to talk about what a line is for example? You don't have a blackboard and you can't mail diagrams.

You said that we could do every part of math. You are wrong.

After the three pre-algebra books, after two courses in algebra (beginning and advanced algebra), comes *Life of Fred: Geometry Expanded Edition.*

In Chapter 1, we do points and lines. In Chapter 2, angles, in Chapter 3, triangles, . . . , in Chapter 12, solid geometry (cubes and pyramids, etc.), in Chapter 12½, geometry in four dimensions.

Wait! I was right. All that stuff needs diagrams. You draw lines and triangles and point to them.

That's true. But in the last chapter of the book, we show how to redo all of geometry *without diagrams.* A 19-year-old kid named Robert L. Moore figured this out in 1899.

In regular old traditional geometry we have statements like *between any two distinct points there exists a line.*

In the geometry invented by Moore, you don't have to have to draw any points or lines.

No other high school geometry textbook ever mentions Moore and his approach to geometry.

Chapter Twenty   Measuring Mass

Second part: the 𝔐ixed 𝔅ag: a variety of problems from this chapter and previous chapters

114. You are planning to spend a day at the library and will need some snacks. You bake 4 identical pizzas and toss them into your backpack along with 7 pounds of beef jerky. The whole thing weighs the same as if you had packed 48 pounds of beef jerky.
   How much does one of your pizzas weigh?

   That's what a word problem might look like in beginning algebra.
   Since you are trying to find the weight of one of your pizzas, you let x equal the weight of one of your pizzas. That's pretty logical, isn't it?

   I picture a giant balance scale.

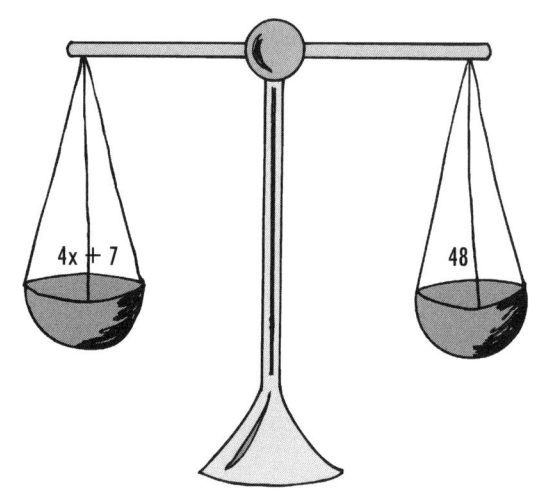

   The 4 pizzas plus 7 pounds would balance with 48 pounds.
In algebra you would write the equation 4x + 7 = 48.
Solve that equation and we will find the weight of one of your pizzas.

454. You have 48 pounds of food in your backpack. The backpack itself weighs 12 pounds. Even without a calculator, you know the whole thing weighs 60 pounds.
   That's too much to carry. You drag it down the street toward the library. It takes 20 pounds of force to drag it at 4 mph.
   As you are dragging it, you reach into your backpack and take out 3 pounds of beef jerky and eat it. How much force will it now take to drag it at 4 mph?

Chapter Twenty-one
A Second Way to Measure Mass

First part: Problems from this chapter

313. Everyone knows that it takes 19 pounds of force to drag your 57 pound backpack at 4 mph down the street toward the library. (You just figured that out in the previous problem.)
    How much force is needed to drag it at 8 mph down the street?

675. If you were dragging it at 4 mph and then later you were dragging it at 8 mph, it took an extra bit of force to get it moving at the faster speed. That extra bit of force is called _____.
    fill in one word

809. You are happily dragging your backpack down the street at 8 mph. Your kid sister grabs it and your speed slows to 7 mph, then 6 mph, then 5 mph, then 4 mph. The force that she was exerting on your backpack is called _____.
    fill in one word

922. Note the skid marks her shoes are making. Her shoes are sliding on the sidewalk. The force (inertia) she is exerting on your backpack times the distance she slides can be measured in foot-pounds.
    Foot-pounds measures work. Work is a transfer of energy. Into what form of energy is she transferring the energy as she slows you down? (The nine forms of energy are chemical, electrical, heat, height, light, motion, nuclear, sound, and spring.)

Chapter Twenty-one   A Second Way to Measure Mass

Second part: the 𝔐ixed 𝔅ag: a variety of problems from this chapter and previous chapters

198. When you kid sister first grabbed your backpack you were going 8 mph. One second later you were going 7 mph. Two seconds after she grabbed the backpack you were going 6 mph.
   Plot (0, 8), (1, 7), (2, 6) and draw a line to estimate when you, your backpack, and your sister will come to a stop.

421. You let go of your backpack and head off to the library. Without your four pizzas and your beef jerky, you figured that you would have to starve at the library.
   You sister emptied out your backpack. She wasn't interested in the food you had packed. She wanted the backpack. To make it hers, she painted it pink.

   After adding 3 coats of pink paint and sticking her 8-pound doll inside, the backpack weighed 17 pounds more than when it was empty.
   How much did each coat of paint weigh?
Hint #1: You start this kind of problem by letting x equal the thing you are trying to find out. In this case, let x = the weight of each coat of paint.
Hint #2: The three coats of paint will weigh 3x.
Hint #3: After applying the paint and sticking her doll inside, the weight will have increased by 3x + 8.
Hint #4: We are told that the weight increased by 17.
Hint #5: 3x + 8 = 17

598. You arrived at the library. At the library snack bar you buy 5 egg sandwiches and a 3-pound onion. The whole thing weighs 10 pounds. How much does each egg sandwich weigh?

Chapter Twenty-one   A Second Way to Measure Mass

756. You can eat 20% of an egg sandwich in a minute. (Translation: In one minute you can eat 0.2 egg sandwiches.)

How long will it take you to eat your 5 egg sandwiches? Use a conversion factor.

844. After twenty-five minutes at the library you decide to look at the books. (We all know what you were doing for the first 25 minutes of your time at the library!)

You find a whole bunch of books by Prof. Eldwood that you like:

Modern Methods of Alligator Removal, 1843
Cooking in the Kitchen of Your Bedroom, 1850
Lost in the City without a Backpack, 1848
What It Means Not to Be the Youngest Sibling, 1842*

These four books plus the 3-pound onion weigh 28 pounds. All the Eldwood books weigh the same. How much does each Eldwood book weigh?

954. In Prof. Eldwood's *Modern Methods of Alligator Removal,* he tells how to make alligator soup.

He wrote: First you make a big pot of boiling water. Then you put the alligator on a slide. Then tip up the slide until the alligator slides into the pot.

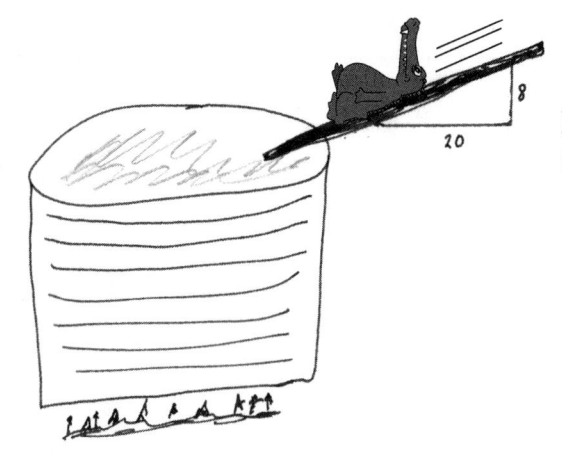

Find the coefficient of static friction, μ_s.

---

* It means that you have either a kid sister or a kid brother or both.

Chapter Twenty-two
Pressure

First part: Problems from this chapter

159. We have had several physics formulas and symbols so far in this book. Change each of these symbols or formulas into words, For example, $P = \frac{F}{A}$ stands for Pressure = $\frac{Force}{Area}$

c

μ

$F = \mu N$

$d = rt$

Hooke's law $F = kx$

μ_k

μ_s

$\mu_s = \frac{rise}{run}$

$W = Fd$ (Hint: think ft-lbs)

$C = \pi d$

284. Your kid sister presses against your arm. She doesn't have a lot of strength. Are you more concerned about (A) the force she applies or (B) the pressure she applies?

600. Right now, suppose you learned that 14 psi (pounds per square inch) of pressure was being applied to every square inch of your skin. Would you want to (A) yell Stop! or (B) would you like that pressure to continue?

Chapter Twenty-two   Pressure

Second part: the 𝔐ixed 𝔅ag: a variety of problems from this chapter and previous chapters

366. In Prof. Eldwood's *Cooking in the Kitchen of Your Bedroom*, he tells his readers that one of the big advantages is that after you take something out of the oven, you have very few steps to make to take it to your bed.

You take your freshly cooked pizza from the oven to your bed. It took 8 steps (also known as 8 paces). How far is it from your oven to your bed? Use a conversion factor. (Your paces are 28".)

452. How far is that in feet? How far is that in feet and inches? *Discussion: Finding how far it is in feet is old stuff. You just use a conversion factor.*

But converting inches into feet and inches when it doesn't come out evenly may need a little review. Suppose I want to convert 135 minutes into hours and minutes. It will look like this . . .

$$\begin{array}{r} 2 \text{ R } 15 \\ 60 \overline{)135} \\ \underline{120} \\ 15 \end{array}$$
  2 hours, 15 minutes

610. For dessert after the pizza, nothing beats having some big red strawberries. To keep this book inexpensive I've done several things: The cover is laminated so that chocolate will be easy to wipe off; the binding is Smyth-sewn, which lasts a lot longer than adhesive binding; the text is in black and white rather than in full color. It will last through several generations of readers rather than having to buy a new book for each reader. That's why you have been asked not to write in this book. The only exception would be if you bought this book with your own money. Then feel free to color these strawberries red.

Big strawberries might be 3 cm (centimeters) across. The strawberries you have are 260% larger than that. How big are they?

754. You put 5 of these big strawberries on a 0.7 kg (kilogram) plate. The whole thing (strawberries + plate) weighs 3.7 kg. How much does each strawberry weigh?

920. How many *pounds* does each of those strawberries weigh? Use a conversion factor. 1 kg ≈ 2.2 lbs.

59

Chapter Twenty-three
Swim Masks

First part: Problems from this chapter

162. Fred's swim mask is in the shape of an ellipse. In the book we learned that its area is 10 square inches (in²). Suppose we know that b is 2 inches. What is the length a? (Use $\pi \doteq 3$ and $A_{ellipse} = \pi ab$.)

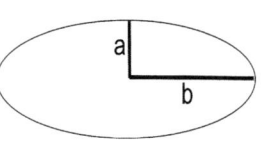

286. One foot = 12 inches.
One square foot = 144 square inches. $1 \text{ ft}^2 = 12^2 \text{ in}^2$

One meter = 100 centimeters. $1 \text{ m}^2 =$ _____ cm^2
fill in a number

488. If Fred's swim mask were rectangular and the width was 2½ inches, what would be the length? Do your work in fractions for this problem. Recall, the area of the mask is 10 in².

571. 1 kilometer (km) = 1,000 meters (m) = 10^3 m.
One million dollars = \$1,000,000 = \$$10^?$

592. Which is larger: 10^2 or 2^{10} ?

875. 1 cubic foot = 12^3 cubic inches. $1 \text{ ft}^3 = 12^3 \text{ in}^3$
1 cubic yard = x^3 cubic inches. What does x equal?

986. A point has no dimension. •
A segment has one dimension.   —————— It has length.
A square has two dimensions.   It has area.

A cube has three dimensions.   It has volume.

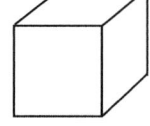

Four dimensions?   Possible or not possible?

60

Chapter Twenty-three   Swim Masks

Second part: the 𝔐ixed 𝔅ag: a variety of problems from this chapter and previous chapters

120. This is a triangle where the two bottom angles are equal. The top angle is equal to 30°. You know that the sum of the three angles of any triangle is equal to 180°. Find the equation and then solve it.

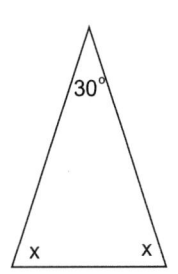

314. A **right triangle** is a triangle that contains a right angle. All right angles are 90° angles.

Suppose we have a right triangle in which the medium-sized angle is twice as large as the smallest angle.

What is the size of the smallest angle in that triangle?

659. I draw a whole bunch of triangles, and I measure all the angles. In every case I find that the sum of the angles in each of the triangles I've drawn is equal to 180°.

From this bunch of measurements I conclude that the sum of the angles in any triangle is equal to 180°. Is this an example of inductive or deductive reasoning?

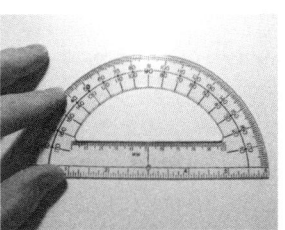

My handy protractor with which I measure angles

900. The volume of a ball (called a **sphere** in mathematics) is

$$V_{sphere} = (4/3)\pi r^3$$

where r = the radius of the sphere.

The radius of a sphere is the distance from the center to the surface.

The diameter of the earth is roughly 8,000 miles. What is the volume of the earth? (Let $\pi = 3$ for this problem.)

61

Chapter Twenty-four
Density

First part: Problems from this chapter

121. If dv = w, find the value of d.
(Translation: Solve the equation for d.)

316. Solve each of these equations.

$5 = \dfrac{x}{3}$

$\dfrac{y}{9} = 6$

$\dfrac{w}{7} + 2 = 10$

535. You have an 8-foot tall stack of pizzas beside your bed. Each pizza has an area of 75 square inches. Somehow the stack got shoved slightly to the right. You were looking for someone to blame. It couldn't be your kid sister since you don't allow her near your pizzas. It couldn't be the wind since your window is shut. Maybe you bumped the stack when you were making your bed. (You *do* make your bed, don't you?) 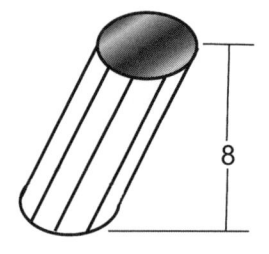
You straighten the stack up. Does the volume decrease, increase, or stay the same?

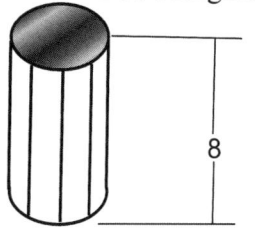

674. What is the volume of your stack of pizzas?

901. If the radius of your stomach is 3", what is its volume? Let's suppose your stomach is in the shape of a sphere. $V_{sphere} = (4/3)\pi r^3$. (For this problem, let $\pi = 3$.)

Chapter Twenty-four   Density

Second part: the 𝓜ixed 𝓑ag: a variety of problems from this chapter and previous chapters

258. Solve 376 = 3x + 5x

499. Suppose we have a triangle in which all three angles are equal. What is the size of each of those angles?

605. Solve 3x + 7 + 9 = 19

713. We did a lot of equations in the *Your Turn to Play* in this chapter in *Life of Fred: Pre-Algebra 0 with Physics*, page 110. The last equation (in problem 4) was p = dh. What did each of these three letters stand for?

870. You have pizzas and strawberries next to your bed. There is only one more thing that you need. You install a giant barrel filled with chocolate milkshake. The faucet is 7 feet below the top of the barrel. What is the pressure at the faucet?

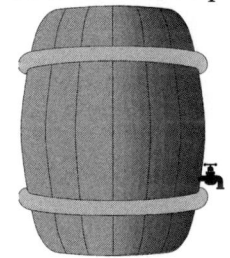

(The density of good quality chocolate milkshake, in which you use lots of cream, is 50 pounds/ft³.)

910. If your barrel of chocolate milkshake were in the shape of a cylinder, with straight sides instead of curved sides, it might have these dimensions in feet.

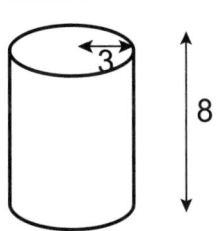

A) What is the area of the circle that has a radius equal to 3?
B) What is the volume of this cylinder? (Use Cavalieri's principle.)
C) What is the weight of the chocolate milkshake in this container?

(Hint for problem C: We know the density from the previous problem. We know the formula from the first problem in the *Your Turn to Play*.)

Chapter Twenty-five
Why Bubbles Float Upward

First part: Problems from this chapter

450. The buoyancy of an object in water is dv. Translation: It is buoyed up by the density of water times the volume of the object.

    Suppose your kid sister didn't like the fact that you were enjoying your pizza while you were lying in bed in your own bedroom. She ran in, grabbed your pizza (volume = 0.2 cubic feet), ran down the hallway, and tossed it into bathtub (where your mom was taking a bath).

    Your mom objected to having to share her tub with a pizza (but didn't object to your kid sister having stolen a pizza from you).

    What is the upward force (the buoyancy) of your pizza? (For this problem assume the density of water is 62 lbs./ft^3.)

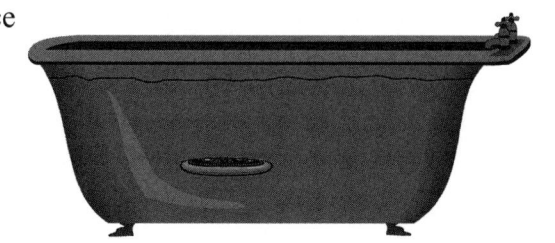

493. You have just determined the buoyancy of your pizza. If your pizza weighs 18 pounds. How hard will it be pressing against your mom's legs?

573. Does the buoyancy of the pizza in the bath water depend on how deep the pizza is below the surface?

676. Does the buoyancy of the pizza in the bath water depend on how dense your pizza is?

Chapter Twenty-five   Why Bubbles Float Upward

Second part: the 𝔐ixed 𝔅ag: a variety of problems from this chapter and previous chapters

210. Your mom lifted that pizza off her legs and up to the surface of the water. That was a distance of 18 inches. In the next problem we are going to compute the work she did in foot-pounds. In this problem convert 18 inches into feet using a conversion factor.

356. How many foot-pounds of work did she do in raising that pizza? (Back in problem #493 on the previous page, we found that the weight of the pizza in water was 5.6 pounds.)

495. In raising that pizza she converted what form of energy into what other form of energy? The nine forms of energy are chemical, electrical, heat, height, light, motion, nuclear, sound, and spring.

710. "Here. Take this pizza back to where you got it," your mom shouts. Your kid sister holds out her little chubby hand and your mom plops the 18-pound pizza on it.

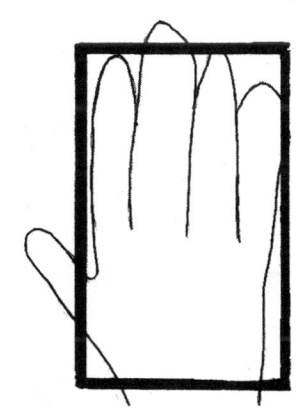

Her hand has the same area as a 2"×3" rectangle.

What is the pressure on her hand? (This is a two-step problem.)

847. She takes the pizza back to your bedroom. You are busy carrying a couple of boxes containing your CD collection to your bed. She puts your pizza on top of the boxes. It does not slide.
    Then she gives it a little push and the pizza slides down the box on onto your lap. This does not feel good. (This is litotes—for those readers who remember what that means.)
    Does this show that $\mu_s < \mu_k$?

65

Chapter Twenty-six
Why Things Sink

First part: Problems from this chapter

190. Your lap is now cold and wet. You do not have nice feelings toward your sister. (litotes)
   You set down your boxes containing your CD collection and head to the bathroom to clean off your pants.

Meanwhile, she opens one of the boxes, takes your favorite CD, and walks over to your aquarium.

She says, "Okay fishes. Do you want to hear some Fred music?" She throws it into the tank.

Does it float or sink?

The data:
✓ Density of water is 62 lbs./ft^3, which equals 0.57 oz./in^3. We worked that out on page 129 in *Life of Fred: Pre-Algebra 0 with Physics*.
✓ The disc with its cover measures 5" × 5½" × ¼".
✓ The CD with its cover weighs 4 ounces.

340. "You threw my favorite compact disc into my aquarium! Why did you do that?" you ask your little sister with the chubby hands.
   That's a silly question to ask a three-year-old. They sometimes do things with very little thought. Their brains aren't that developed yet.
   You explained to her that you don't want to reach down to the bottom of the fish tank to get your CD. She smiled and says, "I fix that." She got out a big bag of salt and poured it into the tank. The density of the water became 0.59 oz./in^3. Will your CD float to the top of the water?

Chapter Twenty-six   Why Things Sink

Second part: the 𝔐ixed 𝔅ag: a variety of problems from this chapter and previous chapters

405. You take your *Best of Fred* CD out of the water and dry it off. CDs aren't hurt by a little water. Your kid sister wants to help. She takes one of your dead fish and dries it off. You are incredulous.*

    She asks in her little three-year-old voice, "Aren't you going to help?" and hands you to other dead fish so that you can dry it off.

    She smiles and says, "Can I have your fishy tank? You don't have any use for it anymore."

    You say nothing, and she guesses that means yes. She drags it across your bedroom floor and out into the hall. It takes 8 pounds of force to push ~~your~~ her tank across the floor at 4 mph. The tank weighs 24 pounds. Find either the coefficient of static friction or the coefficient of kinetic friction or find both.

496. Your big dead fish weighs three times as much as your little dead fish. Together they weigh 13 ounces. How much does your little dead fish weigh?

    (The first step in any word problem is to let x = the thing you are trying to find out.)

642. You lift that pair of fish 24 inches upward and carry them to the wastebasket. How many foot-pounds of work did you use in lifting them?

757. You carry that pair of fish (13 ounces) clutched to your chest as you transport them to their final resting place. They stay 4 feet off the ground. You walk 11 feet to the wastebasket. How much work would a physicist say that you had done in carrying those fish?

951. You decide to sing your *Fishy Goodbye Song* as you walk toward the wastebasket. You start with the lung muscles and wind up with song. Describe how energy is converted among the nine forms of energy: *motion/heat/light/sound/electrical/height/nuclear/spring/chemical.*

---

* Incredulous = you can't believe it. (in-CRED-you-lus, where CRED rhymes with *bread*.)

Chapter Twenty-seven
Nose and Brain

First part: Problems from this chapter

213. You were only mildly curious about what you kid sister was going to do with the fish tank that used to be yours. You figured that as long as she didn't bother you, she could do anything she wanted with that tank.

A minute later you hear strange sounds coming from her bedroom. *Whahee! Ho de hum! Zippee!* You are now being bothered, and you are more than mildly curious. You head down the hall to investigate.

The kid has put on her bathing suit and is pretending to be Esther Williams* swimming in the fish tank. The density of that salt water is 0.59 oz./in^3. She could float in that water. What can you say about the density of your kid sister? This question is not super easy. (litotes)

500. Just as kids need toys in the bathtub, your three-year-old sister needed a "play toy" in her new swimming pool. She threw her 2-pound pillow into the water. That's a 32-ounce pillow. It floated. The density of that salt water is 0.59 oz./in^3. What is the volume of the part of the pillow that is underwater? Round your answer to the nearest cubic inch.

614. The pillow began to be soaked with water. Its density increased. Its volume remained the same. Its ____?____ increased.
fill in one word

758. The pillow will begin to sink when its density exceeds what number?

---

* Esther Williams? Ask your grandparents. They might remember her. She was born in 1921 and starred in many "aquamusicals" that featured her swimming and diving. She was very popular. In 1945 and 1946 and 1947 and 1948 and 1949, she was in movies that were in the top 20 movies for that year. Come to think of it, maybe your should ask your great-grandparents about Esther.

Chapter Twenty-seven   Nose and Brain

Second part: the 𝔐ixed 𝔅ag: a variety of problems from this chapter and previous chapters

165. She got out of the tank after a couple of hours. Her fingers were starting to look like prunes with lots of wrinkles.

It was time to empty that tank. Here was her three-year-old reasoning:
✓ I can't lift the tank and take it to the toilet to empty it.
✓ If I just tip the tank over, it will make a big puddle in my bedroom.
✓ Mommy always asks me to be neat. I will be neat. I will drill a hole in the tank and let the water out. (Hole is 18 inches below the surface of the water. The salt water has a density of 0.59 oz./in^3.)

There were two things that her little mind didn't consider:
① The water would pour out of the hole and make a big puddle in her bedroom.
② Putting a hole in the side of an aquarium will wreck the aquarium. It's hard to fix a hole in glass.

She drilled the hole. What is the pressure of the water at that hole?

215. After she drilled the hole she realized two things:
① The water pours out of the hole and makes a big puddle in my bedroom.
② Putting a hole in the side of an aquarium wrecks the aquarium.
But she had one happy thought:
③ Without all that water the tank will be easier to drag.

It takes only 3 pounds of force to get that 7-pound tank started. Find the coefficient of static friction or the coefficient of kinetic friction or both. (Would you like a little surprise? *No! I, your reader, don't like your "little surprises."* Okay. How about a little hint for this problem? *Now you're talking! I love hints.* The hint is that with what you know in this problem and problem #405 on page 66 of this *Zillions* book, you can find both coefficients. *Hey! That's a surprise.*)

506. She drags the tank back into your bedroom. "Here. Take it. Your tank is no good," she cries. You notice that the hole she drilled was in the shape of an ellipse, with the width being 1" and the height being 0.4". What is the area of that ellipse?

Chapter Twenty-eight
The Long Straw

First part: Problems from this chapter

125. The weight density of air is about 0.08 lbs./ft^3. The weight density of helium is about 0.01 lbs./ft^3.

    In a room full of air, like your bathroom, you weigh, say, 100 pounds. If you take your scale into your kid sister's bedroom, you will notice that the "air" is funny. She has been popping her helium balloons and the room is now filled with pure helium.

    You weigh yourself. Will you weigh more, less, or the same?

Technical note: Your body enjoys having oxygen in its lungs. Your brain especially likes oxygen. It only weighs about three pounds, but it uses about 25% of the oxygen needed by your body. Your brain is an oxygen hog. You wouldn't want to stay in your sister's helium room too long. In much less than a minute you would pass out.

Hold it! Something is wrong here. I, your reader, want to know how your kid sister didn't die in this helium room that she created. That's easy. When she was changing into her bathing suit to go swimming in her aquarium, she looked in her swimming closet. It was filled with swim fins, rubber duckies, swim noodles, goggles, and . . . a scuba tank. When she was popping her balloons, she put on her scuba tank and could continue to breathe oxygen.

487. Of course, the amount of change on your scale would depend on your body's volume, v. The buoyance (upward force) on your body in air is equal to $d_{air}v$. The buoyance in helium would be $d_{helium}v$.

    The weight gain on the scale (from your original 100 pounds) would be found by the difference in buoyancies.
        Your weight gain would be $0.08v - 0.01v$.
        What does $0.08v - 0.01v$ equal?

626. To get practical for a moment, suppose your weight gain was 0.084 pounds. What is the volume of your body?

Chapter Twenty-eight   The Long Straw

Second part: the 𝔐ixed 𝔅ag: a variety of problems from this chapter and previous chapters

678. Actually, the air in her room didn't smell funny. Helium doesn't have any smell. It smells just like air (which is a mixture of nitrogen and oxygen).

You probably would not have noticed a gain of 0.084 pounds on your scale when you weighed yourself in her room

You would not have taken much notice of her wearing a scuba tank. She's three years old. Sometimes she dresses up as princess. Sometimes, as a monkey. Sometimes, as a clown. Last Halloween she dressed up as a ear of corn. She painted her teeth yellow.

You stood there looking at the mess in her bedroom. You told her, "You really ought to clean this stuff up."

She responded, "Ooof! Ooof!" She was using her scuba tank.

You passed out. In ten minutes you would be dead if nothing changed.

The first thing she did was to take your scale. You had left it in her room so she figured it was hers. She lifted that 4-pound scale 16 inches in the air and carried it into her bathroom. How much work (in foot-pounds) did she do?

719. She took off her 6-pound scuba tank and hung it on a giant rubber band. The rubber band stretched 8 inches. Then she took ~~your~~ her scale and added it to the rubber band. How far did it stretch?

721. And, of course, without the scuba tank she passed out. Your mom had gotten out of the bathtub and put on a robe. Then she thought to herself *It's too quiet!* She ran to where the noise wasn't and realized what your kid sister had done.

She popped your kid sister on top of you and dragged the pair of you out into the hallway where there was fresh air. If μ_k was 0.54, what would it have been if she had just dragged you?

71

Chapter Twenty-nine
Nature Loves Vacuums

First part: Problems from this chapter

264. Your kid sister's eyes popped open before yours did. Why?

368. Your mom gave your kid sister a big vacuumy* kiss on her cheek. Part of her cheek bulged out slightly where she was being kissed.

 Everyone knows that vacuums do not apply force. (At least people who have read Chapter 29 know that.) Where did the force come from that made her cheek bulge?

398. If the atmosphere of earth were pure helium instead of a mixture of oxygen and nitrogen and we repeated the glass tube experiment, what would be the result?

 The glass tube experiment: Take a 50-foot long glass tube, seal one end, fill it with water, stick the open end into a lake. When this is done in an atmosphere of air, the water falls until it is 34 feet above the surface of the lake.

759. Your mom takes her "little dear one" out onto the balcony of your house which overlooks Sluice Lake—a lake filled with *the world's sweetest soft drink*[SM]. Sluice is also the world's most dense soft drink since it is 99% sugar and 1% water.

 She gives your kid sister a 34-foot straw. What is going to happen?

---

* I know *vacuumy* isn't a word. I can't think of the word I need. It would be silly to say *vacuum-filled kiss*. Vacuums do the opposite of filling. I couldn't say *big sucking kiss* because I wanted to include the word *vacuum* in the question.

 How many syllables in the word *vacuum*? If you said two, you would be right. That's the most common pronunciation. If you said three, you would be right. The second most common pronunciation is VAC-you-em.

Chapter Twenty-nine   Nature Loves Vacuums

Second part: the 𝔐ixed 𝔅ag: a variety of problems from this chapter and previous chapters

193. Your eyes open. You see your mom out on the balcony combing your kid sister's hair and trying to comfort her because she couldn't get any Sluice from the straw. Your mom comes down the hall, steps over your body, and heads down to Sluice Lake. She fills a large glass with 2 pounds of Sluice and carries it back to the house, up the stairs, down the hall—again, stepping over your body—and into the little chubby hands of your kid sister.

Your first thought is that she has done 68 foot-pounds of work (= 34 feet × 2 pounds). You note that part of the time she was carrying that weight down the hall. Was your first thought correct?

344. Your dad comes down the hall. He doesn't step over you. He asks, "Hey. Why are you laying there?"

You think to yourself *Should I correct his English and tell him I'm lying here, not laying here? To lie is to recline. To lay means to lay something. You lay eggs. You lay your backpack down.*

Then you think *I must have passed out. The last thing I remember is taking my scale into my sister's bedroom and weighing myself. Her room was a mess with lots of popped balloons on the floor.*

Upon hearing no response from you after waiting 0.39 seconds, he steps over you and joins his wife and "little dearest" daughter out on the balcony.

Enough! you think to yourself. *It's time to do a little reasoning.*
Ever since Chapter 6 I have seen a lot of things:
Observation #1: She (your kid sister) removed all the chicken from the soup and no one told her not to. (Chapter 6)
Observation #2: She stole one of the drawers from my chest of drawers and no one told her to give it back. (Chapter 7)
Observation #3: Her alligator bit my arm and no one kissed it or put a bandage on it. (Chapter 8)
Observation #4: In Chapter 9 she shouted at me, "Nothing but the best for my 'gator" and later shouted, "I just made a plate of food for my 'gator," and no one told her to use her inside voice and not shout. Every time I raised my voice even a little bit, my parents hushed me up.

73

Chapter Twenty-nine   Nature Loves Vacuums

Observation #5: She stole my whole chest of drawers and no one mentioned the Eighth Commandment. (Chapter 10)

Observation #6: Also in Chapter 10 she dragged a dirty garden hose into the house and squirted me when I was in my own bedroom. No one told her that she wasn't supposed to do that.

Observation #7: In some families they don't let their kids bounce their big rubber balls in the hallway. (Chapter 11)

Observation #8: In no other family that I know of do they let their kid's alligator make bounce, bounce, bounce, bounce, bounce, bounce, bounce sounds at 11 at night. (Chapter 11)

Observation #9: Every time I did anything, she cried. I sang too loudly, she cried. I don't apologize to her alligator, she cried. I don't read her favorite book to her, she cried. Mom just let her cry even though it drove me crazy. (Chapter 12)

Observation #10: Nobody objected when she put her stupid alligator on the kitchen table when I'm trying to eat. (Chapter 13)

Observation #11: When I wanted to protect my box of Wizard shredded cereal by taking it into the living room, mom shouted at me. (Chapter 13)

Observation #12: When I headed into my bedroom to eat some pizza in quiet, she switched my pizza for a Jolly Joker Joking Pizza and laughed at me when I tried to bite it. This was a very cruel thing for her to do[*]. That hurt a lot.

Observation #13: She painted my bedroom door. Why did they let her get away with that? (Chapter 14)

Observation #14: In Chapter 15 when I complained to my dad, he just said, "That's not bad work for a three-year-old." Whose side was he on?

Observation #15: She watched stupid three-year-old stuff on television with the volume turned up so loud that it drove me nuts. When I asked her to turn it down a bit, mom yelled at me. Mom pulled me down the hall and into my bedroom. (Chapter 17)

Observation #16: I was dragging my 57-pound backpack down the street and she hung onto it, slowing me down. No one disciplined her. (Chapter 21)

Observation #17: She stole my backpack and painted it pink. (Chapter 21)

---

[*] There's an old saying: *You can mess with my car. You can mess with my cell phone. You can mess with my chest of drawers. But never mess with my pizza.*

Chapter Twenty-nine   Nature Loves Vacuums

Observation #18: She stole one of my freshly baked* pizzas, and mom didn't make any fuss about that. (Chapter 25)

Observation #19: She shoved the wet pizza on my lap. (Chapter 25) She dropped my favorite CD into my fish tank. (Chapter 26) She dumped salt into my tank and killed my fish. (Chapter 26) She stole my fish tank. (Chapter 26) No parent offered any discipline.

Observation #20: She cut a hole in my tank and made a big puddle of water in her bedroom. She then dragged that useless tank back into my bedroom and cried, "Your tank is no good." She's claimed it's my fault. No one stuck up for me. (Chapter 27)

Observation #21: She made a mess with popped balloons in her bedroom and no one told her to clean it up. (Chapter 28)

Observation #22: In Chapter 28 she stole my scale, but no one told her that was wrong. (I can't blame her for accidentally almost killing me.)

Observation #23: She got a big kiss. She was taken out to the balcony. She was called "little dear one." She got a Sluice straw. She got her hair combed. (Chapter 29) I was ignored.

Observation #24: My dad said six words to me and then headed to look after his "little dearest daughter."

After all this thinking, you come to the conclusion that your parents care more for your kid sister than they do for you.

My questions for you: After 24 observations, are you 100% certain that this is true? Are you using deductive reasoning?

---

* *Freshly baked pizzas* doesn't have a hyphen in it. But *ten-inch sandwich* does.

Freshly modifies *baked*. *Ten* modifies *inch*. What's the difference?

The rule in English is that if that first word ends in *-ly* then you omit the hyphen.

    newly created painting
           hastily constructed house
                plainly marked path
    easy-going father
           well-kept bedroom
                worth-while use of your time

Chapter Thirty
Weighing the Atmosphere

First part: Problems from this chapter

458. It is often inconvenient to erect a 35-foot tall glass tube in your bedroom, fill it with water, and notice that the water falls to 34 feet. On the other hand, that is one way to measure the air pressure.

And knowing the air pressure can have a real benefit.

I, your reader, can't imagine what that would be. Telling your friends that the air pressure is 2,123 lbs./ft² or 14.74 lbs./in² really won't make you that popular.

You are right. In fact, in case your mother never told you, inserting numbers in your conversations with friends is often not a good idea. Do you think you would be more popular if you . . .

- brag that you own 6 cell phones
- announce that you are worth $500,000 or
- declare that you have had 8 colds this last year?

Now, where was I? Yes, knowing air pressure can have a real benefit. That's because air pressure—atmospheric pressure—can change slightly. Very often, when the atmospheric pressure is falling, that indicates a low pressure weather pattern is coming. **RAIN** or **SNOW**. When the atmospheric pressure is rising, **SUNSHINE** may be on the way.

If you use Sluice instead of water, the glass tube doesn't have to be as long because Sluice is denser than water. But a glass tube that is 15 or 20 feet tall still is hard to erect in your bedroom.

We need a liquid that is really dense. How about a metal that is liquid at room temperature? Mercury is the only one.

The density of mercury is about 0.49 pounds per cubic inch. It's heavy! How many inches of mercury is needed to create a pressure of 14.7 lbs./in²?

The formula relating pressure, density, and height was the answer to the **BIG QUESTION** in Chapter 24.

Chapter Thirty   Weighing the Atmosphere

Second part: the 𝕸ixed 𝔅ag: a variety of problems from this chapter and previous chapters

619. Your kid sister said that she wanted some crayons to do a little coloring. Your mom ordered 6 cases of crayons. It came in a box that weighed 0.2 pounds. The whole thing weighed 47 pounds. How much did one of the cases weigh?

   You always start by letting x equal the thing you are trying to find out. Super hint: Look for the question mark in the problem.
   Let x = the weight of one of the cases of crayons.
   Then 6x = the weight of the six cases.
   0.2 = the weight of the packaging.
   47 lbs. is the weight of the whole thing.

At this point you can write the equation. The hardest part of these kind of problems (called word problems) is getting from the English to the equation. Solving the equation is fairly easy. Students often want to leap from the English directly to the equation. They want to skip writing the "Let x = . . . " and the "Then . . . " statements.
   You might be able to do it here in these very simple word problems, but in algebra where things get a bit more complicated, you may die in your attempt to avoid writing out these four or five lines. Now is the time to develop good habits.*

725. Graph the points (2, 1) and (5, 3). What is the slope of the line that connects those two points?

921. From the balcony of your house your kid sister couldn't suck up Sluice using a 34-foot long straw. She could have if it had been water, since water is less dense than Sluice.
   Your mom tries to comfort her by saying, "If we were now at our vacation home that is near the pass that goes into Yosemite (elevation 10,000 feet), then you could use this 34-foot straw to suck up Sluice."
   Does this sound true to you?

---

* The other good habit is learning to write legibly. It doesn't have to be super neat, but poor handwriting can lead to a lot of what my mother used to call careless mistakes.

Chapter Thirty-one
Water Fountains

First part: Problems from this chapter

167. Your flip a light switch in your house.

The light goes on.

Electrical energy is changed into what two other forms of energy?

300. A circuit is a journey that goes around and comes back to where it started.
✧ Electrons make a circuit.
✧ Horses on a race track.
✧ Earth around the sun.
  Name at least a couple more examples.

503. Physicists (at least today) think that all electrons in the universe are all alike. The happy little electron ☺ who leaves the negative end of the battery is virtually identical to the happy little electron ☺ who has made the trip through the light bulb and is arriving back at the positive end of the battery.

That's not true of horses who have run as fast as they can around a race track.

It's not true of the earth after it has made a year's trip around the sun. Each year the earth grows older in a thousand ways.

Each morning when you wake up, you are different than the previous morning. You are not a happy little electron ☺. Name two ways that you are different.

Chapter Thirty-one   Water Fountains

Second part: the 𝔐ixed 𝔅ag: a variety of problems from this chapter and previous chapters

644. This felt strange. You had never known that there was Sluice Lake right below the balcony of your house. You get up from the hall floor and dash down the stairs and out the front door. You head around the house to the side where the balcony is.

There, right behind the trees, was Sluice Lake. The density is 0.8 pounds per cubic inch. If you dove 10 feet under the surface of the lake, what would be the pressure?

761. You spot this duck in a boat. That didn't bother you. At this point you figured that *anything* is possible. You had never noticed a lake near your house. The lake was filled with Sluice. In this *Alice in Wonderland* world, a duck with a hat driving a motorboat was almost ordinary. Captain Duck was going 8 mph (miles per hour). How long would it take him to go 300 feet. (5,280 feet = 1 mile) Use conversion factors.

923. Ducks normally just paddle around in a lake. Then you notice that this lake has something special in it. Now you understand why Captain Duck likes to ride in a motorboat.

This floating alligator is your kid sister's. What is the volume of the Sluice that it displaces?

So you don't have to turn back to previous pages in this book, we learned in Chapter 8 that the alligator weighs 18 pounds, and in this chapter that the density of Sluice is 0.8 lbs./in^3.

976. Solve $3w + 61 = 15w + 1$

79

Chapter Thirty-two
A Small History

First part: Problems from this chapter

164. Luigi Galvani held down that frog with brass hooks. Brass is an alloy (mixture) of copper and zinc. Common brass is 63% copper and 37% zinc.

If Luigi had 2 pounds of common brass hooks, what is the weight of the copper in those hooks?

Of course, common brass isn't the only brass there is. There is admiralty brass, aich's alloy, aluminum brass, arsenical brass, cartridge brass, DZR brass, delta metal, free machining brass, gilding metal, high brass, leaded brass, lead-free brass, low brass, manganese brass, muntz metal, naval brass, nickel brass, nordic gold, prince's metal, rose brass, tombac, tonval brass, and yellow brass.

My favorite is prince's metal. It is 75% copper and 25% zinc. It looks a lot like gold but is much cheaper.

The brass section of a band, which includes the trombone, tuba, cornet, trumpet, baritone horn, euphonium, and the French horn, contains . . . brass. (You could have guessed that.) Horns made out of brass sound better than those made out of lead. And brass has a good resistance to corrosion. Translation: A good resistance to spit.

Brass is mentioned 32 times in the book of Exodus. (I counted them in my concordance.)

216. Metals conduct electricity much better than non-metals. That is why the wires in your house are made of copper rather than rubber or plastic. Wires are often coated with rubber or plastic so that if you touch them, the electricity won't pass from the copper to your body.

Silver is a better conductor of electricity than is copper. Why aren't the wires in your house made of silver?

Chapter Thirty-two   A Small History

Second part: the 𝔐ixed 𝔅ag: a variety of problems from this chapter and previous chapters

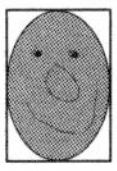

333. You head back to your bedroom and shut the door. The fact that the door has a big ugly ellipse painted on it doesn't seem to bother you as much as it had.

You don't have a chest of drawers anymore. On the floor is 4 pounds of socks and underwear. You lift them up 30 inches and put them on a shelf. How many foot-pounds of work was that?

460. Also on the floor is your copy of Christina Rossetti's poetry. You pick it up, read some lines from her poem "Later Life"

Love pardons the unpardonable past:
Love in a dominant embrace holds fast
His frailer self. . . .

and put the book on that shelf with your socks. You did 5 ft-lbs. of work to put the book on the shelf. How heavy is that book?

620. Also on the floor is your collection of eleven colored pencils and the 9-gram pencil jar. You put the pencils in the jar and put it on the shelf with your socks and your Christina book. The pencils and the jar weigh 75 grams. How much does each pencil weigh?

724. After you have loaded all the things that were in your chest of drawers onto that shelf, the shelf starts to tip. The first thing to start to slide is your Christina Rossetti book. Find the coefficient of static friction between the book and the shelf.

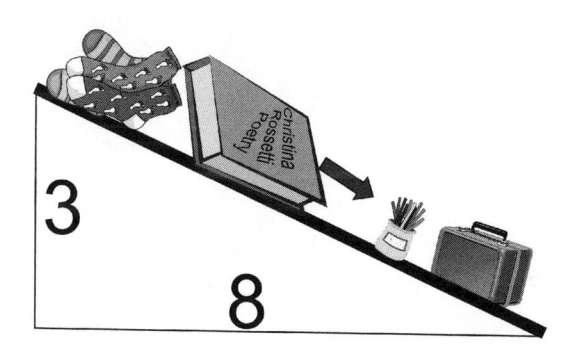

Chapter Thirty-three
Kitty Café

First part: Problems from this chapter

110. Here is a schematic with four different things in it. I'll number them as #1, #2, #3, and #4.

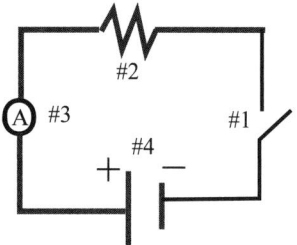

You know that #2 is a resistor. It could be a lamp, an electric mixer, or an electric chair.

What do #1, #3, and #4 stand for? They can be found among the items in this list:

★ a battery or other power source
★ a 1957 Chevrolet car
★ an ammeter, which measures the flow of electrons
★ a switch
★ a big building at night

217. 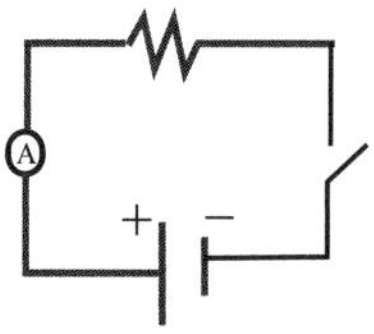 In this diagram, Ⓐ is an ammeter, which measures how many electrons are flowing past that point in the circuit.

Which is true?
A) There are zillions of electrons flowing through the ammeter.
B) There is no way to tell how many electrons are flowing through the ammeter.
C) There are zero electrons flowing through the ammeter.

461. The same question as the previous problem but using this schematic.

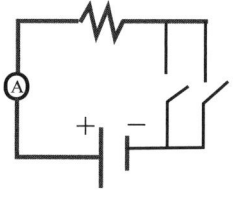

Chapter Thirty-three Kitty Café

Second part: the 𝔐ixed 𝔅ag: a variety of problems from this chapter and previous chapters

504. How about now?

A) There are zillions of electrons flowing through the ammeter.

B) There is no way to tell how many electrons are flowing through the ammeter.

C) There are zero electrons flowing through the ammeter.

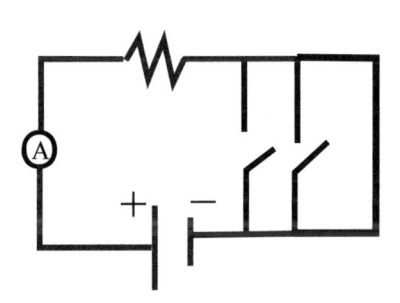

658. Your jar of pencils is sitting on that shelf that has tipped. It is not moving. Copy this diagram and do a little art work by resolving the weight of the jar, W, into a normal force, N, and a frictional force, F.

745. Since we know that the jar is not moving, we know that the coefficient of static friction, μ_s, between the jar and the shelf is greater than what number? (This takes us back to Chapters 13 in which Kingie found a formula that flabbergasted Fred.)

846. Your kid sister comes into your bedroom (without knocking). She takes three pencils out of your pencil jar. Does this make your jar more or less likely to start sliding?

927. With her little chubby hands, she moves the jar a little bit when she is stealing your pencils. Does this make the jar more likely to start sliding?

952. She says, "Don't you think my mom did a nice job of combing my hair?" You wish she had said "*our* mom," but she didn't. She leaves your bedroom, slamming the door on the way out. The energy of motion is changed into sound energy.

You walk over to the door, and she suddenly opens the door. The doorknob hits you with 28 pounds of force. She says, "I want all those pencils." You will have a blue bruise tomorrow. Do you wish that the area of that doorknob was larger or smaller than it is now?

Chapter Thirty-four
Volts and Amperes

First part: Problems from this chapter

166. If ammeter #1 reads 45 A, and ammeter #2 reads 15 A, what will ammeter #3 read?

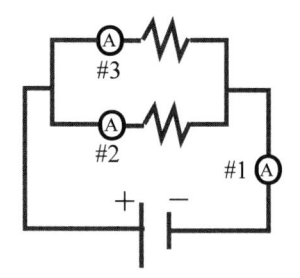

240. If #1 reads 12 amps, #2 reads 3 amps, and #3 reads 6 amps, what will #4 read?

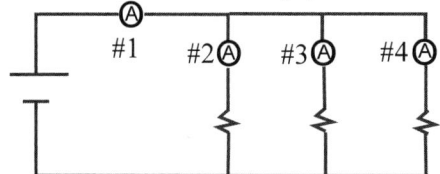

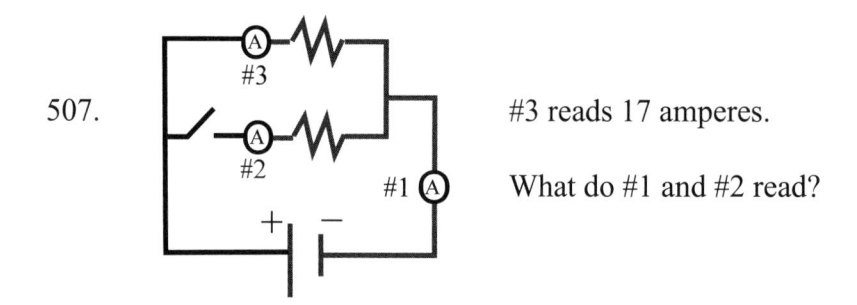

507. #3 reads 17 amperes.

What do #1 and #2 read?

588. One ampere represents a flow of 6.24×10^{18} electrons per second. How many electrons per minute is that? (Use a conversion factor.)

Chapter Thirty-four   Volts and Amperes

Second part: the 𝓜ixed 𝓑ag: a variety of problems from this chapter and previous chapters

679. In this circuit a 9-volt battery will create a current of 10 amperes.

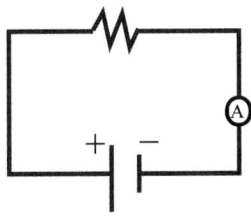

If I replace the battery with a 110-volt power source, what will the current be?

(Use a conversion factor.) Round your answer to the nearest amp.

726. She takes all your colored pencils but leaves the broken jar on the floor. You are not surprised.

You grab a broom and push the 0.08 pounds of the mess 12 feet across the room and into a corner so that no one will step on it. The coefficient of friction between the mess and the floor is 0.5. How much work did you do?

848. You are eating some rubbery spaghetti. That will take your mind off of the mess in the corner of your room. If you hang a 2-gram meatball on that strand of spaghetti, it stretches 6 mm (mm = millimeters. Ten mm equals one cm. One cm is a little less than half an inch.)

If you hang a 3-gram meatball, it stretches 9 mm. If you hang a 4-gram meatball, it stretches 12 mm.

Graph these three points. (2, 6,), (3, 9), (4, 12)

928. $3x - 5 = 2x$ and $30 = 5y + 9$
    Is it true that $x < y$? $<$ means "less than"

Chapter Thirty-five
Ohms

First part: Problems from this chapter

259. There are several nice things to say about Tradition. One pleasant thing about traditions is that you know what to expect. wE START SENTENCES WITH CAPITAL LETTERS. We shake hands with our ʇɥƃıɹ hands.

The post office doesn't encourage you to be creative when you put a stamp on an envelope.

But following tradition does have its costs. The hero in "Fiddler on the Roof" had to go through much heartache to follow the traditions of his world.

If you were a normal kid, you put up a big fuss when your parents insisted that you not go to the bathroom in your pants. But now that you follow the adult tradition, you have a big advantage in dating over those who don't.

In physics, the tradition is _?_ is the letter that stands for resistance and _?_ is the letter that stands for rate of flow of the current. Eight volts is written as 8 V. Eight ohms is written as 8 _?_ .

265. Copy this diagram and label the appropriate parts with 24 V, 2 Ω, and 12 A.

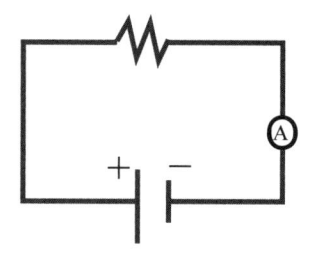

Chapter Thirty-five   Ohms

Second part: the Mixed Bag: a variety of problems from this chapter and previous chapters

459. #1 reads 15 A.
#2 = 10 Ω.
#4 = 15 Ω.
What will ammeter #3 read?

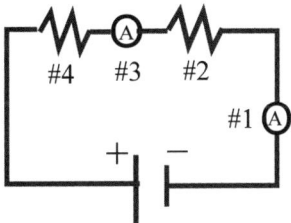

505. You think about your dream house that you will have when you are older. It will have a long driveway and many rooms. Everyone seems to like having lawns that are rectangular.

But that would be boring.

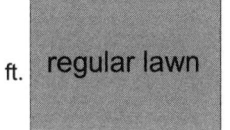

To make your house unique, you want your lawn in the shape of an ellipse.

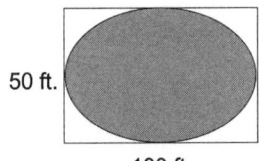

What is the area of a "regular lawn"?
What is the area of your elliptical lawn?
(Use $\pi = 3$ for this problem.)

680. (continuing the previous problem) Your gardener will plant your elliptical lawn. It will cost $40 for each 250 square feet he plants. What will be the total cost? (Use a conversion factor.)

764. Your kid sister will have a house that has 3,000 square feet in it. Your house will be 260% larger than hers. How many square feet will be in yours? (Hint: The answer is not 7,800 ft^2.)

818. Your kid sister won't mind that your house is much larger than hers. She plans on making lots of visits to your house to "borrow" things that you have, such as colored pencils. Now you know why your dream house has a long driveway (see problem #505). If your driveway is 1,200 feet long and she can walk at 3 feet per second, how long will it take for her to walk up your driveway? Give you answer in minutes and seconds.

Chapter Thirty-six
Ohm's Law

First part: Problems from this chapter

127. Three guys helping each other push is like three batteries in series. If each guy could push 20 pounds, then their total push would be 60 pounds.

This [1.5 V 1.5 V 1.5 V batteries in series] could be replaced by a single battery with what voltage? [? V]

270. The same thing is true of resistors in series.

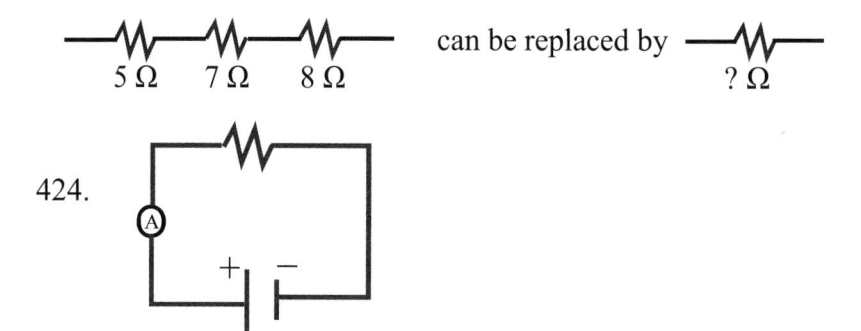

[5 Ω, 7 Ω, 8 Ω in series] can be replaced by [? Ω]

424. [schematic: ammeter, resistor, and battery in a loop]

In this schematic . . .

Question A) If $I = 7$ A and $R = 8$ Ω, what is the voltage of the battery?

Question B) If $R = 16$ Ω and $V = 8$ V, what does the ammeter read?

Question C) If $V = 110$ volts and $I = 37$ A, what will the ammeter read?

661. Ohm's law is $V = IR$, where V is the voltage, I is the amperage, and R is the resistance. Which of these are true?

1) $V = RI$ 2) $V/R = I$ 3) $R = V/I$ 4) $IR = V$

Chapter Thirty-six   Ohm's Law

Second part: the 𝓜ixed 𝓑ag: a variety of problems from this chapter and previous chapters

681. In the west wing of your dream house are 12 bedrooms. They are all the same size. A painter says that he could paint all the rooms for $5,000. Of that $5,000, $3,524 would be the cost of the paint and the rest is his labor cost. How much would be the labor cost for each of the bedrooms?

740. Your dream house should have a swimming pool. When it was constructed, the workers made a slight mistake.

Instead of water, they filled the pool with OIL.
It's density is 50 lbs/ft^3.
The pool measures 20' × 30' × 8'.
What is the weight of the oil?

765. You don't want the oil to go to waste. You have a rusted combination lock.   You toss it into the pool. It sinks to the bottom.

What is the pressure on that lock?

820. If you went swimming in that oil pool would you be more or less likely to sink than if you were in a pool filled with water. (Density of water is about 62 lbs./ft^3.)

878. It's not very pleasant to swim in oil, but you think of a way to make a lot of money. You advertise in The KITTEN Caboodle newspaper . . .

𝒯𝐻𝐸 𝒦𝐼𝒯𝒯𝐸𝒩 Caboodle

The Official Campus Newspaper of KITTENS University   Friday 10 a.m. Edition 10¢

exclusive
Fred Has Girlfriend!
KANSAS: It couldn't happen to a nicer guy. All of the campus is talking. **(continued on page 17)**

Advertisement: Come swim in my oil pool. It will cure your dry skin! 50¢ per swim

---

If 5,000 come and swim, how much will you make?

89

Chapter Thirty-seven
Parallel Circuits

First part: Problems from this chapter

115. The electric circuit for your pool has two items in parallel: a lamp (100 Ω) and a heater (500 Ω).

When the lamp and the heater are both turned on, what is the total resistance in the circuit?

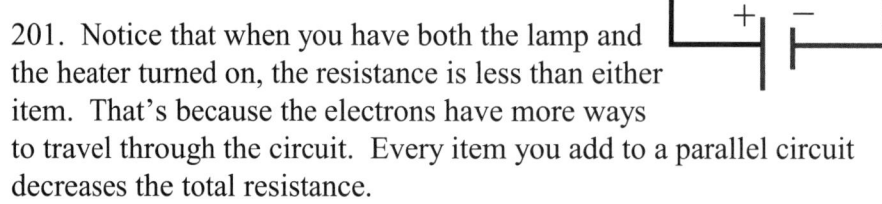

201. Notice that when you have both the lamp and the heater turned on, the resistance is less than either item. That's because the electrons have more ways to travel through the circuit. Every item you add to a parallel circuit decreases the total resistance.

Suppose you have ten 10 Ω lamps in parallel. What would be the total resistance?

281. Ohm's law is V = IR. If you decrease the resistance and keep the voltage the same, what happens to the amperage?

393. You pay the power company for the number of electrons you use.*
One ampere is defined as 6.24×10^{18} electrons per second.

Part A) If you run a 55 Ω light bulb on a 110 volt circuit, how many amperes will be flowing?

Part B) How many electrons will you be using each second?

---

* Technically, you pay for the number of electrons times the force (the voltage) that they are delivered. "Lazy" electrons with a force of, say 10 volts, won't light your lights as well as electrons delivered with 100 volts.

You pay for watt-hours, not electrons. Watts = amperes times volts. W = IV. And to keep the number of zeros down, you are billed for kilowatt-hours, not watt-hours. One kilowatt equals 1,000 watts.

Without these simplifications, your light bill might read: You used 807,000,000,000 000,000,000,000,000 electrons delivered at 110 volts during this last month.

Chapter Thirty-seven   Parallel Circuits

Second part: the 𝔐ixed 𝔅ag: a variety of problems from this chapter and previous chapters

426. You kid sister was one of the 5,000 people who came to swim in your oil-filled pool. She, of course, didn't pay the 50¢.

She dove to the bottom of the pool and picked up the combination lock. 🔒 Having soaked in the oil, it was no longer rusty. In fact, it worked perfectly now. She kept it.

When she was underwater (or rather underoil), her buoyancy was 80 pounds. Since she doesn't weigh 80 pounds, she floated to the surface. What is the volume of your kid sister? (Recall, the density of the oil is 50 pounds/cubic foot.)

445. With her little chubby feet all coated with oil, she could pretend she was skating on the concrete near the pool. She ran and then slid. She shouted, "Wheeeeee!" and came to a stop. μ_k is very low between oily feet and concrete. The energy of motion was converted into what other form of energy?

589. You put out a stack of towels for your guests to use after they have been swimming.

All the towels are the same size (2' × 3'). Your kid sister took all the towels and played with them. She then stacked them but did a sloppy job. The pile was just as high but the towels weren't nicely aligned. The two stacks have the same volume because of the ____?____ principle.

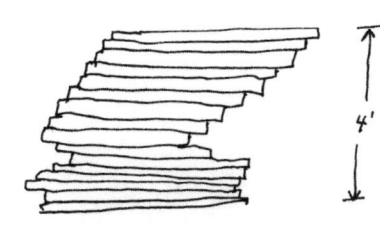

811. Even though this was just your dream house, things were starting to go wrong because of her. (We won't mention names!)

For this last question of this book, I'm going to ask you to do a little creative writing. Finish this story . . .

```
Your kid sister was trying to be helpful.
One of the 4,999 guest was smoking.  She grabbed
that cigarette and. . . .
```

91

104–106 Complete Solutions and Answers

104. Which is larger, a thousand or a million?
 A million. 1,000,000 > 1,000

105. Yes. Recently some physicists are playing with the idea that the speed of light in a vacuum might not always have been a constant. They have a conjecture that earlier in the history of the universe, light might have been going at a different speed.

hunch → conjecture → theory → law

Even when scientists declare some idea to be a law, it isn't 100% certain.

106. Change 7 faradays into ampere-hours using a conversion factor. (In physics handbooks, you can learn that 26.8 faradays equals one ampere-hour.)

The conversion factor will be either $\dfrac{26.8 \text{ faradays}}{1 \text{ ampere-hour}}$ or it will be $\dfrac{1 \text{ ampere-hour}}{26.8 \text{ faradays}}$

I choose the one so that the faraday units will cancel.

$$\dfrac{7 \cancel{\text{ faradays}}}{1} \times \dfrac{1 \text{ ampere-hour}}{26.8 \cancel{\text{ faradays}}}$$

$$= \dfrac{7}{26.8} \text{ ampere-hours}$$

Discussion: What to do with the $\dfrac{7}{26.8}$?

Alternative #1: You could just leave it.

Alternative #2: You could multiply top and bottom by 10 to get rid of the decimal. You would get 70/268

Alternative #3: You could do the long division $26.8 \overline{)7.00}$ and get ≈ 0.26119402985074626865671641791045 . . .

(I'm not making up this 0.26119402985074626865671641791045! If you get out a piece of paper, you can check my work.)

Complete Solutions and Answers 110–115

110. Here is a schematic with four different things in it. I'll number them as #1, #2, #3, and #4.

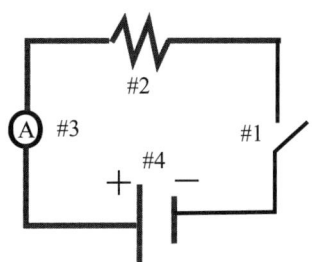

You know that #2 is a resistor. It could be a lamp, an electric mixer, or an electric chair.

What do #1, #3, and #4 stand for?

#1 is a switch
#4 is a battery or other power source
That leaves #3. It could be: ★ a 1957 Chevrolet car or ★ an ammeter, which measures the flow of electrons or ★ a big building at night. (Hint: It is not a Chevrolet or a building.)

114. Solve $\quad 4x + 7 = 48$

Subtract 7 from both sides $\quad 4x = 41$

Divide both sides by 4 $\quad x = \dfrac{41}{4}$

Do the arithmetic $\quad x = 10.25$ pounds or $x = 10¼$ pounds Either way is fine.

$$\begin{array}{r} 10.25 \\ 4\overline{)\,41.00} \end{array} \quad \text{or} \quad \begin{array}{r} 10\,¼ \\ 4\overline{)\,41} \\ \underline{4} \\ 01 \\ \underline{0} \\ 1 \end{array}$$

115. The electric circuit for your pool has two items in parallel: a lamp (100 Ω) and a heater (500 Ω). When the lamp and the heater are both turned on, what is the total resistance in the circuit?

$\dfrac{1}{R} = \dfrac{1}{R_{lamp}} + \dfrac{1}{R_{heater}}$ becomes $\quad \dfrac{1}{R} = \dfrac{1}{100} + \dfrac{1}{500}$

Doing the arithmetic $\quad\quad\quad\quad\quad\quad\quad\quad \dfrac{1}{R} = \dfrac{6}{500}$

Invert the fractions $\quad\quad\quad\quad\quad\quad\quad\quad R = \dfrac{500}{6}$

Doing the arithmetic $\quad\quad\quad\quad\quad\quad\quad\quad R = 83⅓$ ohms

93

| 119–120 | Complete Solutions and Answers |

119. Which came first, m or c?

If you were to go out walking in the woods, I doubt that you would find meters. No bird would chirp, "Hey! Here's a meter."

Meters do not occur in nature. Someone on Mars walking on a Mars desert would not come up with the same length and call it a meter.

But the light on Mars and the light in Kansas share the same c. They both travel at the same speed.*

So you can email your friend on Mars and say, "Measure the speed of light. Now we use that speed to define what a meter is."

c came first. Then m.

120. This is a triangle where the two bottom angles are equal. The top angle is equal to 30°. You know that the sum of the three angles of any triangle is equal to 180°. Find the equation and then solve it.

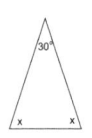

Let x = the measure of one of the bottom angles.

Then since the measure of all three angles is 180°	2x + 30 = 180
Subtract 30 from both sides	2x = 150
Divide both sides by 2	x = 75

Each of the bottom angles is equal to 75°.

(In geometry we will call those bottom angles the **base angles**.)

---

* We are talking about the speed of light in a vacuum. That's the definition of c. When light travels through air or water or glass, it slows down.

Complete Solutions and Answers 121–124

121. If dv = w, find the value of d.
(Translation: Solve the equation for d.)

$$dv = w$$

Divide both sides by v $\quad d = \dfrac{w}{v}$

We have gone from dv = w to d = w/v by dividing both sides by v.

Going the other direction, we could go from d = w/v to dv = w by multiplying both sides by v.

If we wanted to solve $4 = \dfrac{x}{7}$ we multiply both sides by 7 and get 28 = x.

122. Your empty chest of drawers now weighed 9 pounds. It would take 6 pounds of force to get it moving. Find μ_s.

The formula for static friction is $F = \mu_s N$.

We know that N = 9 and F = 6 $\qquad\qquad 6 = \mu_s 9$

Putting the number before the letter $\qquad 6 = 9\mu_s$

Dividing both sides by 9 $\qquad\qquad\qquad \dfrac{2}{3} = \mu_s$

The arithmetic: $\dfrac{6}{9} = \dfrac{2}{3}$ when you divide top and bottom by 3

The coefficient of static friction, μ_s, is $\dfrac{2}{3}$

123. You lift that 7-pound target three feet upward (not upwards) and put it into the garbage can. How much work was done?

W = Fd becomes $\qquad\qquad$ W = 7 × 3 = 21 ft-lbs

124. Find x.   100x = 37

You start with $\qquad\qquad$ 100x = 37

Divide both sides by 100 $\qquad\qquad x = \dfrac{37}{100}$ or 0.37

| 125–126 | Complete Solutions and Answers

125. The weight density of air is about 0.08 lbs./ft^3. The weight density of helium is about 0.01 lbs./ft^3.

In a room full of air, like your bathroom, you weigh, say, 100 pounds. If you take your scale into your kid sister's bedroom, you will notice that the "air" is funny. She has been popping her helium balloons and the room is now filled with pure helium.

You weigh yourself. Will you weigh more, less, or the same?

Your weight doesn't change in going from your air-filled bathroom to her helium-filled room. Gravity has the same attraction in both spots.

But the buoyancy of air is different than the buoyancy of helium.

The upward force (the buoyancy) in your bathroom = (density of air)(your volume). In her bedroom it equals (density of helium)(your volume). Since the density of helium is less than the density of air, the upward force is less. Your scale would now read more than 100 pounds.

If you went swimming in a pool filled with light fluffy water, you would sink to the bottom of the pool and have a really tough time getting to the surface.

"Light fluffy water"? I, your reader, can't imagine that.

How about being thrown into a pool of cotton candy? That stuff would be very nasty to breathe.

126. In the book Fred had passed out and was on top of the broken door that lay on the floor. Kingie had to push with a force of 21 pounds to get the door moving with Fred on it.

The coefficient of static friction, μ_s, between Kingie's shoes and the floor is 0.7.

Kingie is too light to push Fred and the door. He only weighs one pound. How many pounds of art supplies must be put in his pocket to get Fred and the door moving?

$F = \mu N$ becomes $\quad 21 = 0.7N$

Divide both sides by 0.7 $\quad 30 = N$

$0.7 \overline{)21.}$

Kingie weighs one pound. He will have to stick at least 29 pounds of art supplies in his pocket, so that he will weigh at least 30 pounds.

$\dfrac{30}{7.\overline{)210}}$

127. This could be replaced by a single battery with 4.5 volts.

96

Complete Solutions and Answers | 155–156

155. You are walking at a constant speed. The distance you walk is proportional to which of these?

    B) the number of minutes you walk

If you walk for twice as many minutes, you will walk twice as far. If you walk for ten times as many minutes, you will walk ten times as far. The distance you walk is proportional to how long you walk.

156. The noise coming out of her room is driving you nuts. She has these programs going all the time.

    Which of these "facts" is not true?

Fact A: Energy in the form of sound is leaving her bedroom.
Fact B: The Law of Conservation of Energy: In any closed system, the amount of energy cannot change.
Fact C: Since energy (sound) is coming out of her bedroom it is not a closed system.
Fact D: If you have a system in which energy is only leaving the system, then the amount of energy in the system must decrease as time goes on.
Fact E: There is no energy entering her bedroom.

    Certainly, noise is coming out of her room (Fact A). The Law of Conservation of Energy is true (Fact B).* The definition of a closed system is a system in which no energy enters or leaves the system. So Fact C is true. If you have a bucket full of water and the only thing happening is water is being poured out of the bucket, then certainly the amount of water in the bucket will decrease over time. Fact D is true.

    Is there energy entering her bedroom? She's watching television. Every television I know gets plugged into the wall. The electricity is entering the system of her bedroom. Fact E is false.

    I, your reader, want to argue with you. What if she has big batteries that are powering her TV. Then no energy is entering her bedroom. You, Mr. Author, didn't think of that.

    I did. If she had batteries, then energy would only be leaving her bedroom. Eventually, as energy only left her system—remember the water in the bucket example—her room would become quiet. I wrote that "She has these programs going all the time."

---

* We must note that because of $E = mc^2$, in which matter and energy can be interchanged, the amount of energy includes the matter inside the closed system.

| 157–159 | Complete Solutions and Answers |

157. Time in the Imperial system is measured in seconds. Have you ever been told what the unit used for time is in the metric system?

    The answer is yes.*

158. When the ball is heading downward, it has the energy of motion.

159.
c is the speed of light in a vacuum. 299,792,458 meters per second

μ is the constant of proportionality for friction, also known as the coefficient of friction

F = μN   The Force of friction is equal to the coefficient of friction times the Normal force between the two surfaces.

d = rt   Distance equals Rate times Time

Hooke's law F = kx   The weight on a spring, F, is equal to the spring constant, k, times the distance stretched, x.

μ_k the coefficient of kinetic (moving) friction

μ_s the coefficient of static friction

$\mu_s = \frac{rise}{run}$ the coefficient of static friction is equal to

W = Fd   Work is equal to Force times distance

C = πd   Circumference of a circle is equal to pi times the diameter

---

*** Hey! Wait a minute! I, your reader, am not happy with a simple "yes." What in blazes is the unit of time that corresponds to seconds in the Imperial system?**

    I thought you would never ask. In all the couple hundred countries in the world that use the metric system, the unit of time is . . . seconds.

    Stop! This can't be. Then in the metric system we would have 60 seconds = 1 minutes; 60 minutes = 1 hour. That's not how the metric system operates. Where are all the ten times conversions?

    Who ever told you that the metric system is perfect? Over in Japan, Germany, and Brazil they still have 7 days = 1 week.

Complete Solutions and Answers 160–163

160. Simplify $\frac{73\mu}{73}$

This equals µ. When you multiply µ by 73 and then divide that answer by 73, you get the number you started with, which is µ.

161. Solve $\quad\quad\quad\quad\quad\quad\quad$ 8.9 = x + 3.94

Subtract 3.94 from both sides $\quad\quad\quad$ 4.96 = x

The hardest part isn't the algebra. It's the arithmetic.

$$\begin{array}{r} 8.90 \\ -\ 3.94 \\ \hline 4.96 \end{array}$$

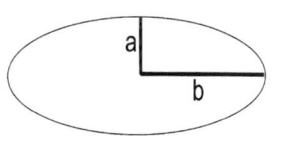

162. Fred's swim mask is in the shape of an ellipse. In the book we learned that its area is 10 square inches (in²). Suppose we know that b is 2 inches. What is the length a? (Use $\pi \doteq 3$ and $A_{ellipse} = \pi ab$.)

$A_{ellipse} = \pi ab$ becomes $\quad\quad$ 10 = 3a(2)
Multiplying $\quad\quad\quad\quad\quad\quad\quad$ 10 = 6a
Divide both sides by 6 $\quad\quad\quad\quad$ 1⅔ = a

The length a is equal to 1⅔ inches.
If you worked in decimals, a would equal $1.\overline{6}$

$$\begin{array}{r} 1.66666 \\ 6\overline{)10.00000} \end{array}$$

163. The alligator target has a diameter of 3 feet. What is the radius of that circle?

A radius is one-half the length of a diameter. One-half of 3 feet is 1½ feet. Or 1.5 feet if you prefer decimals.

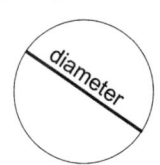

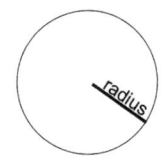

| 164–167 | **Complete Solutions and Answers** |

164. Common brass is 63% copper and 37% zinc. What is the weight of copper in 2 pounds of common brass hooks?

    We want 63% of 2 pounds. We know both sides of the *of* so we multiply.

$$63\% \times 2 \text{ pounds}$$
$$0.63 \times 2 \text{ pounds}$$

Doing the arithmetic   1.26 pounds is the weight of the copper

165. The hole is 18 inches below the surface of the water. The salt water has a density of 0.59 oz./in^3. What is the pressure of the water at that hole?

    In Chapter 22 we answered the 🐦🐦🐦 🐦🐦🐦🐦🐦🐦🐦 of how much pressure is at a depth of h in a liquid of density d. It was p = dh.

    p = dh becomes   p = 0.59 oz./in^3)(18 inches)
    Doing the arithmetic   p = 10.62 oz./in^2

166. If ammeter #1 reads 45 A, and ammeter #2 reads 15 A, what will ammeter #3 read?

    The flow of electrons through #2 and #3 must equal the flow through #1.

    Ammeter #3 will read 30 A.

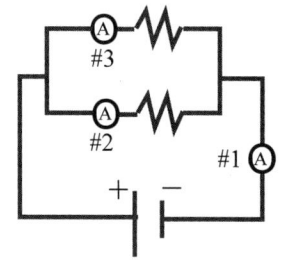

167. Your flip a light switch in your house. The light goes on. Electrical energy is changed into what two other forms of energy?

    Here are the nine forms: *motion/heat/light/sound/electrical/height/nuclear/spring/chemical.*

    Certainly part of the electrical energy is converted into light. If you remember back to the old days when incandescent lights (lights with filaments in them) were more popular, those bulbs got so hot that you couldn't touch them.* Even the modern light bulbs give off some heat. And some old fluorescent bulbs would sometimes hum. So in that case electrical energy was also converted into sound.

---

* To be precise, you could touch them, but you would get burned.

Complete Solutions and Answers | 181–189

181. Is the length of a meter in 1668 the same as the length of a meter today?

No. In 1668 John Wilkins defined the meter as the length of a pendulum with a period of two seconds.

In 1791 they changed the definition to one-ten millionth of the distance from the North Pole to the equator.

In 1889 they changed it again. This time to the length between two scratches on a metal bar.

In 1960 they changed it again. This time they made a definition that you could describe on the telephone to someone on Mars.

✻ This time it didn't depend on the earth's gravity (as the pendulum did).
✻ This time it didn't depend on the shape of the earth (as the North Pole to the equator did).
✻ This time it didn't depend on some metal bar.

With the 1960 definition, the guy on Mars could go get some krypton and make the measurement and know how long a meter was.

189. How much can you stretch that spring so that after you remove the alligator the spring will be good as new?

Once you pass the elastic limit (and enter the plastic region) you have wrecked the spring. It won't come back to its original shape. It will be time to buy a new spring.

You may not remember this. All wristwatches used to have hands. The straps were made from real leather—real leather made from real cows. Stranger still, you could hear your watch. It would go tick—tick—tick—tick. It had no batteries. It was powered by a spring. Every day you had to remember to wind your watch. If you forgot, then it wouldn't go tick—tick—tick—tick.

So that you wouldn't pass the elastic limit on the watch's spring, they designed your tick—tick—tick—tick watch so that it was hard to over wind it.

101

190 Complete Solutions and Answers

190. She throws the 4-oz. CD into the tank. Does it float or sink?
✓ Density of water is 62 lbs./ft³, which equals 0.57 oz./in³.
✓ The disc with its cover measures 5" × 5½" × ¼".
✓ The CD with its cover weighs 4 ounces.

> Best of Fred

The buoyancy of the CD = dv, where d is the density of water and v is the volume of the disc with its cover. We know d. We need to find v.

$v = \ell wh$ where ℓ = length, w = width, and h = height.

$v = 5 \times 5½ \times ¼ = \frac{5}{1} \times \frac{11}{2} \times \frac{1}{4} = \frac{55}{8} = 6\frac{7}{8}$

The buoyancy of the CD = dv = $0.57 \times 6\frac{7}{8}$

I'm going to work in decimals. I think it will be easier than using 0.57 = 57/100

$0.57 \times 6\frac{7}{8} = 0.57 \times 6.875^* = 3.91875$ ounces of upward force.

The downward force is the weigh of the CD, which is 4 ounces.

Since 3.91875 < 4,
    the CD will sink
        to the bottom
             of your aquarium.

---

* You memorized in the previous book, *Life of Fred: Decimals and Percents*, the Nine Conversions:

$\frac{1}{2} = 50\%$ $\frac{1}{3} = 33⅓\%$ $\frac{2}{3} = 66⅔\%$

$\frac{1}{4} = 25\%$ $\frac{3}{4} = 75\%$ $\frac{1}{8} = 12½\%$

$\frac{3}{8} = 37½\%$ $\frac{5}{8} = 62½\%$ $\boxed{\frac{7}{8} = 87½\%}$

These nine are probably the most useful ones to learn by heart.

I also carry around in my head $\frac{1}{6} = 16⅔\%$ and $\frac{5}{6} = 83⅓\%$.

102

Complete Solutions and Answers | 192–194

192. It takes 100 pounds of force to push your 242-pound rabbit across the kitchen floor. You weigh 120 pounds. What must the coefficient of friction, μ, be between your shoes and the kitchen floor in order for you to push that rabbit?

You want to be able to generate 100 pounds of force. Your weight is 120 pounds. $F = \mu N$ becomes $100 = \mu 120$

Divide both sides by 120 $\quad\quad \dfrac{100}{120} = \mu$

Reducing the fraction $\quad \dfrac{100}{120} = \dfrac{50}{60} = \dfrac{5}{6} \quad\quad \mu = \dfrac{5}{6}$

Actually, if μ were exactly equal to $\dfrac{5}{6}$ then both you and the rabbit would both slide. The super technically correct answer would be $\mu > \dfrac{5}{6}$

193. Your first thought is that she has done 68 foot-pounds of work (= 34 feet × 2 pounds). You note that part of the time she was carrying that weight down the hall. Was your first thought correct?

Work is defined as Fd, where F is the force (weight) and d is the change in vertical distance of that weight.

You and I would think that carrying a weight down a hall would involve work, but physicists are only looking at a change in vertical distance. Your first thought was correct.

194. If it takes 4.2 pounds to get it moving, what is the coefficient of static friction between that glob and the floor?

$F = \mu N$ becomes $\quad\quad 4.2 = \mu_s 6$
Putting the number before the letter $\quad 4.2 = 6\mu_s$
Dividing both sides by 6 $\quad\quad 0.7 = \mu_s$

$$\begin{array}{r} 0.7 \\ 6\overline{)4.2} \\ -4\,2 \\ \hline 0 \end{array}$$

The coefficient of static friction is 0.7

| 198–201 | Complete Solutions and Answers

198. Plot (0, 8), (1, 7), (2, 6) and draw a line to estimate when you, your backpack, and your sister will come to a stop.

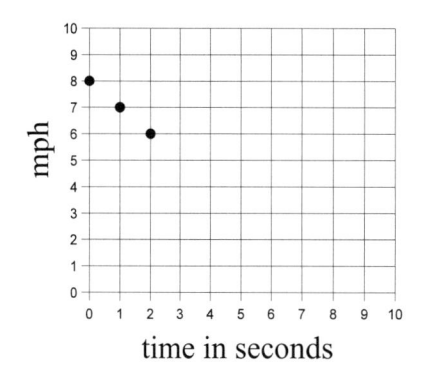

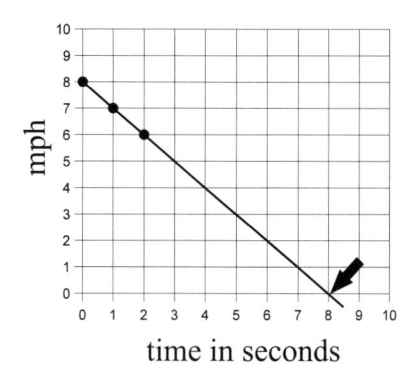

My best estimate is that you will come to a stop (mph = 0) at 8 seconds.

200. How many degrees is ∠1?

In △ ABC we know that ∠B is 60°.
We know that ∠C is 90°.
We know that all three angles must add to 180°.
So ∠1 must be 30°.   60° + 90° + 30° = 180°

201. Suppose you have ten 10 Ω lamps in parallel. What would be the total resistance?

$$\frac{1}{R} = \frac{1}{R_1} + \frac{1}{R_2} + \frac{1}{R_3} + \frac{1}{R_4} + \frac{1}{R_5} + \frac{1}{R_6} + \frac{1}{R_7} + \frac{1}{R_8} + \frac{1}{R_9} + \frac{1}{R_{10}}$$

becomes $\frac{1}{R} = \frac{1}{10} + \frac{1}{10} + \frac{1}{10} + \frac{1}{10} + \frac{1}{10} + \frac{1}{10} + \frac{1}{10} + \frac{1}{10} + \frac{1}{10} + \frac{1}{10}$

Doing the arithmetic $\frac{1}{R} = \frac{1}{1}$

Inverting both fractions   R = 1

The total resistance is equal to 1 Ω.

Complete Solutions and Answers | 208–212

208. Which is the larger numeral, 6 or 8?

A numeral is what we write down when we are thinking of a number. Certainly 8 is a larger *number* than 6, but 6 is a larger *numeral* than 8.

209. Your elliptical pizza is nicely cooked in your wood-fired oven. You stab it with a BBQ fork and drag it out of the oven (at a constant speed). If you know the pizza weighs 6 pounds, what else would you need to know in order to determine the constant of kinetic friction, μ_k?

If we want to find μ_k in the formula $F = \mu_k N$ and we know that N (the weight) is 6 pounds, what we would like to know is the value of F, which is the force needed to drag the pizza at a constant speed.

210. Convert 18 inches into feet using a conversion factor.

$$\frac{18"}{1} \times \frac{1'}{12"} = \frac{18}{12} \text{ feet} = \begin{cases} 1\frac{1}{2} \text{ feet} & \text{if you like to work in fractions} \\ \text{or} \\ 1.5 \text{ feet} & \text{if you like to work in decimals} \end{cases}$$

211. $A = \pi ab$.
Area, A, is 48 square inches and b is 8.
For this problem we will let $\pi = 3$.
How long is a?

Putting these numbers into $A = \pi ab$, we get	$48 = (3)a(8)$
Multiplying	$48 = 24a$
Divide both sides by 24	$2 = a$

212. We know that on earth 1 kg weighs (approximately) 2.2 lbs. For your pizzas you need a lot of pepperoni. You order 18 kg of pepperoni from Uruguay. How many pounds of pepperoni have you ordered? Use a conversion factor.

$$\frac{18 \text{ kg}}{1} \times \frac{2.2 \text{ lbs.}}{1 \text{ kg}}$$

```
    18
  × 2.2
    36
   36
   396    39.6
```

≈ 39.6 pounds of pepperoni.

213–214 Complete Solutions and Answers

213. The density of that salt water was 0.59 oz./in³. She could float in that water. What can you say about the density of your kid sister?

That she can float means that the buoyancy of the water (upward force) is greater than the downward force of her weight. If she were completely underwater, she would float back up to the surface.

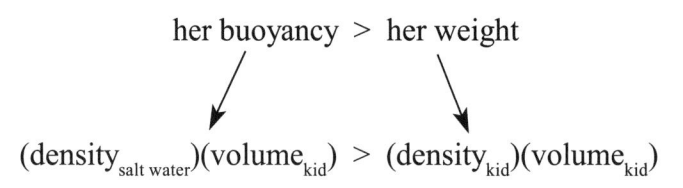

$$\text{her buoyancy} > \text{her weight}$$

$$(\text{density}_{\text{salt water}})(\text{volume}_{\text{kid}}) > (\text{density}_{\text{kid}})(\text{volume}_{\text{kid}})$$

The left arrow	The right arrow
buoyancy = dv	w = dv
is true from Chapter 26.	is true from Chapter 24 in which we defined
	$\text{density} = \dfrac{\text{weight}}{\text{volume}}$

If we divide both sides of $(\text{density}_{\text{salt water}})(\text{volume}_{\text{kid}}) > (\text{density}_{\text{kid}})(\text{volume}_{\text{kid}})$ by volume$_{\text{kid}}$ we get

$$\text{density}_{\text{salt water}} > \text{density}_{\text{kid}}$$

So your kid sister has a density less than 0.59 oz./in³.

In general, things will float if their density of less than the liquid they are in. They will sink if their density is greater than the liquid they are in.

214. You know that N is 240 pounds. For example, if μ were equal to 0.2, what would F equal?

$$F = \mu N = (0.2)(240)$$

```
   240
 × 0.2
  ────
  480      48.0 pounds
```

If μ were equal to 0.2, it would take 48 pounds to push or pull your sack of carrots along the sidewalk.

Complete Solutions and Answers | 215

215. It takes only 3 pounds of force to get that 7-pound tank started. Find the coefficient of static friction or the coefficient of kinetic friction or both. The static friction coefficient, μ_s, is easy to find.

$F = \mu_s N$ becomes $\qquad 3 = \mu_s(7)$
Putting the number in front of the letter $\quad 3 = 7\mu_s$
Dividing both sides by 7 $\qquad\qquad \dfrac{3}{7} = \mu_s$

Finding the coefficient of kinetic (moving) friction is even easier. $\mu_k = 1/3$. We found that in problem #405. **Wait! Stop! You can't do that! We we computed that μ_k it was for a tank full of water. This tank now is almost empty and it's got a hole in its side.*** Recall what μ_k means. It is a measurement of the friction between two surfaces—in this case, between the floor and the bottom of the aquarium. Taking the water out of the tank or drilling a hole in the side of the tank does not change which two surfaces are in contact.

Ha! You just make an English error. You should have written: Taking the water or drilling the hole <u>do</u> not change. . . . That's called subject–verb agreement.

With *and* instead of *or* you would be correct. Taking the water and drilling the hole <u>do</u> not change.

With subjects connected by *or* things are a bit more complicated.

Usual case: If both subjects are singular, then the verb is singular.

An apple or a peach <u>is</u> always in good taste.

Weird case: If the subject closest to the verb is plural, then the verb is plural.

An apple or a six grapes <u>are</u> always tasty.

Six grapes or an apple <u>is</u> always tasty.

In Questions: Always use the plural verb with subjects connected by *or*.

<u>Are</u> Fred or Kingie coming to the picnic?

The Moral: Be a math major rather than an English major. Your life will be a lot easier.

---

* Note the difference between *it's* and *its* in this sentence. *It's* is an abbreviation for *it is*. *Its* is called the possessive form, like his or hers.

| 216–218 | Complete Solutions and Answers |

216. Silver is a better conductor of electricity than is copper. Why aren't the wires in your house made of silver?

  A pound of silver costs about $230 (at today's prices).
  A pound of copper costs about $2.30.
You couldn't afford the house if it had silver wiring.

  Computer manufacturers use silver rather than copper. A better conductor means less resistance, which in turn, means less heat is generated. The amount of silver used is so small that it doesn't significantly increase the price of the computer.

217. In this diagram, Ⓐ is an ammeter, which measures how many electrons are flowing past that point in the circuit.

  Which is true?
A) There are zillions of electrons flowing through the ammeter.
B) There is no way to tell how many electrons are flowing through the ammeter.
C) There are zero electrons flowing through the ammeter.

  Note that the switch is open. There can't be an electrons flowing in that circuit. The correct answer is C).

218. List three things that scientists have proven to be 100% certain.

  Here is a complete list of everything that scientists have proven to be 100% certain:
   1.
   2.
   3.

Complete Solutions and Answers | 240–259

240. If #1 reads 12 amps, #2 reads 3 amps, and #3 reads 6 amps, what will #4 read?

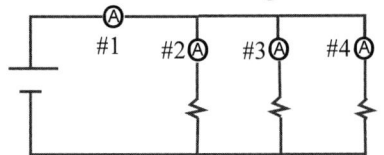

The flow through #2, #3, and #4 must equal the flow through #1.

3 amps + 6 amps + #4 = 12 amps

The flow through #4 must be 3 amps.

251. You want to get from one cliff to the other. The easiest way is to attach a rope to a blimp that is stationed right above the two cliffs and swing from one cliff to the other. Swinging from the left cliff to the right took 12 seconds (and was a lot of fun). . . .The article said that the period was 12 seconds. Was that correct?

Newspapers often get things wrong. A period of a pendulum is the time it takes for the pendulum to make the full trip back and forth. Your period is 24 seconds.

258. Solve $376 = 3x + 5x$

Combine 3x and 5x $376 = 8x$

Divide both sides by 8 $47 = x$

3x + 5x equals 8x
just like 3 toothbrushes plus 5 toothbrushes equals 8 toothbrushes.

259. In physics, the tradition is R is the letter that stands for resistance and I is the letter that stands for rate of flow of the current. Eight volts is written as 8 V. Eight ohms is written as 8 Ω .

| 260 | Complete Solutions and Answers |

260. What is the area of a circle with radius equal to 1½ feet?

In problem 490 in Chapter 9, we started with the area of an ellipse and found the area of a circle.

You point out that circles are just ellipses in which the a and b distances are the same. If you know that $A_{ellipse} = \pi ab$, then what must be the area of a circle?

If a and b are both equal to r, then $A_{ellipse} = \pi ab$ becomes $A_{circle} = \pi rr$

In algebra, we are going to abbreviate rr by r^2. $A_{circle} = \pi r^2$

$A_{circle} = \pi r^2$ becomes $A = \pi(1½)^2$
Which means $A = \pi \times 1½ \times 1½$
Doing the arithmetic $A = \pi \times (2¼)$

$$1½ \times 1½$$
$$= \frac{3}{2} \times \frac{3}{2}$$
$$= \frac{9}{4}$$
$$= 2¼$$

$$4)\overline{9} \quad \begin{array}{c} 2 \; R1 \end{array} = 2¼$$

If you had done the work in decimals, you would have gotten $A = 2.25\pi$ square feet.

You haven't been told to use $\pi = 3$ or $\pi = 3.1$ or $\pi = 2\frac{1}{7}$ or $\pi = 3.1415926535897932384626433832795$ so just leave your answer as $A = 2.25\pi$ square feet.

Complete Solutions and Answers | 261–263

261. To fix the broken door, the carpenter submitted this estimate:

Labor	1,800.00
Materials	$ 186.92
7% sales tax on materials	?

How much was the sales tax?

7% of $186.92 means 7% × 186.92 which is $13.08.

Reminder: Please do not use a calculator in this book.

```
   186.92
 ×   0.07
  130844     13.0844
```

13.0844 rounds off to $13.08.

262. The chicken soup weighed 25 pounds. If μ between the bottom of the bowl and the table was equal to 0.2, how much force would be needed to slide the bowl toward your kid sister at the rate of 6 inches per second?

The speed doesn't make any difference. If you were pushing it at 6 inches per second and at 500 mph, the force needed to push at a constant speed is the same. (Of course, a soup bowl coming at your kid sister at 500 mph might make a big difference to her!)

F = μN becomes F = (0.2)(25) = 5 pounds

```
    25
 × 0.2
    50    5.0
```

You might have done this problem using fractions rather than decimals. $0.2 = 20\% = \frac{1}{5}$

$$(0.2)(25) = \frac{1}{5} \times 25 = \frac{1}{5} \times \frac{25}{1} = \frac{1}{\cancel{5}_1} \times \frac{\cancel{25}^5}{1} = 5$$

263. If the alligator hops on top of the glob, will μ_s increase, decrease, or stay the same? The alligator weighs 18 pounds.

μ_s measures the friction between two surfaces. In this case, it is between the glob and the floor. The alligator on top of the glob doesn't change the two surfaces that are in contact. μ_s doesn't change. What does change is the effort it will take to drag the glob.

264–267 Complete Solutions and Answers

264. Your kid sister's eyes popped open before yours did. Why?

I know this isn't a math problem, but a reading comprehension problem, which is one of the things they test to find out how well you are prepared for college.

The facts: You had passed out. She stole your scale, headed to her bathroom, hung her scuba tank and your scale on a giant rubber band. Then she passed out. Her brain had been oxygen deprived for less time than yours.

265. Copy this diagram and label the appropriate parts with 24 V, 2 Ω, and 12 A.

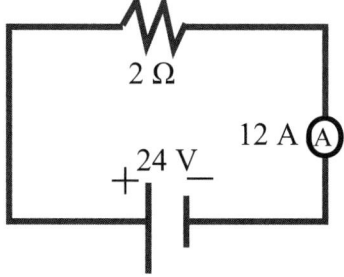

266. These are ordered pairs: (3, 7), (988, 16), (1.008, 62).
Why is this not an ordered pair? (8, 10, 16)

That's easy. A pair means two.

267. If you buy 5 ounces of chocolate and you buy 5 ounces of silver, which weighs more? One regular ounce = 28.349527 grams. Silver (and gold) are measured in troy ounces. One troy ounce = 31.103481 grams.

Since troy ounces are heavier than regular ounces, 5 ounces of silver weigh more than 5 ounces of chocolate.

112

Complete Solutions and Answers | 270–281

270. The same thing is true of resistors in series.

—W—W—W— can be replaced by —W—
 5 Ω 7 Ω 8 Ω   20 Ω

280. Will a change in gravity will affect the period of the pendulum?

The first thing that might come to mind is when Jane held onto you (three questions ago) when you swung from the right cliff to the left. Adding the extra weight did not affect the period.

So it might seem that changing the strength of gravity wouldn't affect the period of a pendulum.

But first impressions are sometimes wrong.

Take a pendulum in, say, of a grandfather clock into outer space where there is almost no gravity. You pull the pendulum to the left . . . and it stays there. It doesn't move. In a million years the pendulum would still be on the left.

Gravity does have an effect on the period of a pendulum. The French Academy was right.

281. Ohm's law is $V = IR$. If you decrease the resistance and keep the voltage the same, what happens to the amperage?

We want to look at I, the amperage. Divide both sides of $V = IR$ by R and we get $\frac{V}{R} = I$.

Now we can look at I when we mess with R.

Let's look at some sample numbers. Suppose V is 100 V and R is 20 Ω. Then $\frac{V}{R} = I$ becomes $\frac{100}{20} = I$. So I is 5 A.

Now let's decrease R to 10 Ω. $\frac{V}{R} = I$ becomes $\frac{100}{10} = I$. So I is now 10 A. The amperage has increased.

When you take a fraction like $\frac{V}{R}$ and decrease the denominator, you increase the value of the fraction.

284–286 Complete Solutions and Answers

284. Your kid sister presses against your arm. She doesn't have a lot of strength. Are you more concerned about (A) the force she applies or (B) the pressure she applies?

    If she were to push your arm with 100 pounds of force, that might shove you down the hallway, but it wouldn't hurt anything except your pride. (The thought of being pushed around by your kid sister might be embarrassing.)

    However, if she wants to play nurse and make you the patient, you might not like even three pounds of force against your arm if she is holding a needle!

    It is pressure that is the important thing.

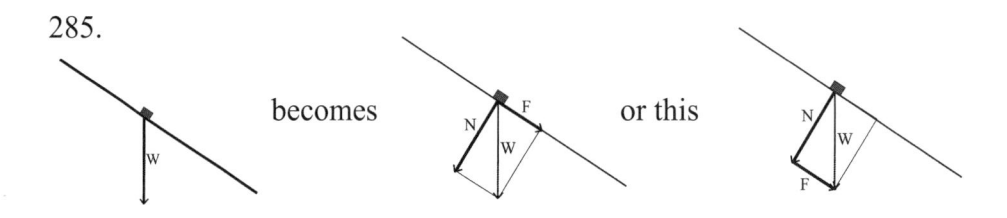

285.

Either one is correct. You can take your choice.

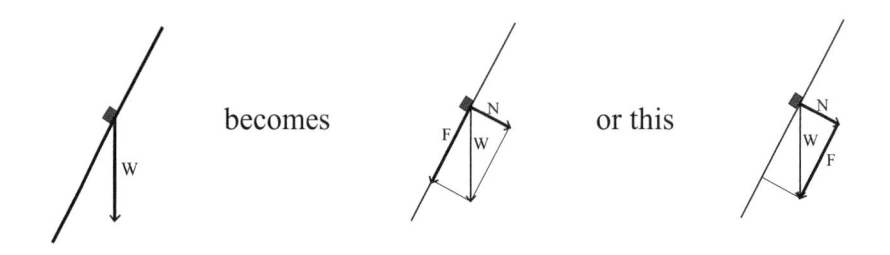

286. One foot = 12 inches.
One square foot = 144 square inches. $1 \text{ ft}^2 = 12^2 \text{ in}^2$

One meter = 100 centimeters. $1 \text{ m}^2 = \underline{100^2} \text{ cm}^2$ or 10,000 cm^2

Complete Solutions and Answers | 287–289

287. It was fun to play in your kitchen in your bedroom. You timed yourself. You walked into your kitchen, and in 5 minutes you could make 6 waffles. On another day you found that you could make 8 waffles in 6 minutes. Plot these two points (5, 6) and (6, 8).

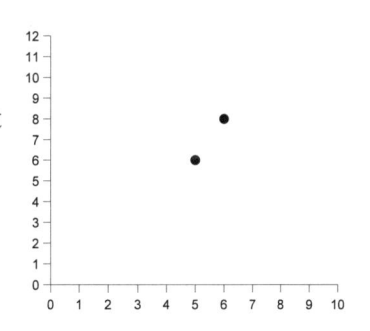

288. You have a chest of drawers. It is 5.1 feet tall. The area of the front is 14.28 square feet. What is the width?

The area of a rectangle is $A = \ell w$.
We know that A is 14.28 and that ℓ is 5.1.
$A = \ell w$ becomes $14.28 = 5.1w$

Divide both sides by 5.1 $\quad \dfrac{14.28}{5.1} = w$

$$5.1 \overline{) 14.28}$$

$$\begin{array}{r} 2.8 \\ 51 \overline{) 142.8} \\ -102 \\ \hline 408 \\ -408 \\ \hline \end{array}$$

Small reminder: Please do not use a calculator in this book. At this point in your education is the perfect time to really learn the basic arithmetic skills.

Simplify $\quad\quad 2.8 = w$

The width of the chest of drawers is 2.8 feet.

289. Your mom pulled you down the hallway and into your room. The work of her pull (from the chemical energy stored in her body into motion) was then transformed into what other form of energy?

The work needed was because of the friction between your body and the hallway floor. Friction transfers motion into heat. That's why you rub your hands together to make them warmer.

115

| 300 | Complete Solutions and Answers |

300. A circuit is a journey that goes around and comes back to where it started. Name at least a couple more examples of circuits.

⋄ Night and day followed by night and day followed by night and day followed by night and day followed by night and day. . . .

⋄ For some people who are paid on the first of each month, the circuit is: Get paid, party it up during the first week of the month, money runs out two-thirds of the way through the month, eat peanut butter for dinner during the last week of the month—and then repeat the cycle.

⋄ For Fred who has been teaching for years at KITTENS University, the cycle begins on the first day of class in the fall. Many new freshmen who are excited about their first time at college. Midterm exams in October at which point some students realize that at real colleges students are expected to study without having to have weekly quizzes to prod them. (Some colleges, unfortunately, are more like warmed over high schools.) Then the spring semester where many of those students who didn't study are no longer enrolled. (They have gone back to live with their parents or are working in fast food joints.) Then summer. Then the cycle begins again on the first day of classes in the fall.

Discussion: There are two ways to view life: one is the cyclical (circuit) and the other is the straight line. Farmers who plant seeds in the spring or women who are pregnant with their eighth child are very aware of how things repeat themselves.

For those who look at the bigger picture, the straight line view seems to be more apparent. Countries are born. They have their early adventurous years. They take their place on the world stage. Some then spread their power and build an empire. They overextend and collapse. Some completely disappear.

Taking the long view, your life is straight line. Most people are born (understatement). They make mistakes as kids. They become adults. They make more mistakes. Some learn why love is important. They get old. They die.

Complete Solutions and Answers 311–312

311. Here are the dimensions of the safe.
What is the area of the front of the safe?

The area is (5)(1.9).

```
    1.9
  ×  5
  ─────
    95
```
9.5 square centimeters
or 9.5 sq cm

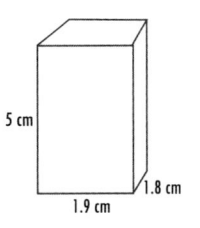

312. You pull the garbage can down the hallway at 3 mph. It weighs 20 pounds and takes 8 pounds of force to pull it.

As you pass your kid sister's door, she opens the door and drops a 40 pound package of used Halloween candy into the garbage can.

How much work will you do in dragging the can 15 feet down the hallway?

(The previous question was a one-step problem. This question is a three-step problem.)

Step one. Find μ_k. $F = \mu_k N$ becomes $\quad\quad 8 = \mu_k 20$
Put the number in front of the letter $\quad\quad\quad\quad\quad\quad 8 = 20\mu_k$
Divide both sides by 20 $\quad\quad\quad\quad\quad\quad\quad\quad\quad \frac{2}{5} = \mu_k$

Step two. Find the force needed to push the 60-pound garbage can down the hallway.

We note that μ_k has not changed. It is still $\frac{2}{5}$ because the same two surfaces are involved.

$F = \mu_k N$ becomes $\quad\quad\quad F = \frac{2}{5} \times 60$
Doing the arithmetic $\quad\quad\quad\quad\quad F = 24$ pounds

Step three. Find the work involved in a force of 24 pounds over a distance of 15 feet.

$W = Fd$ becomes $\quad\quad\quad\quad W = 24 \times 15$
Doing the arithmetic $\quad\quad\quad\quad\quad W = 360$ ft-lbs

| 313–314 | Complete Solutions and Answers

313. It takes 19 pounds of force to drag your 57 pound backpack at 4 mph down the street.

How much force is needed to drag it at 8 mph down the street?

It still takes 19 pounds. If you are dragging it at any constant speed, the force is the same.

314. A **right triangle** is a triangle that contains a right angle. All right angles are 90° angles.

Suppose we have a right triangle in which the medium-sized angle is twice as large as the smallest angle.

What is the size of the smallest angle in that triangle?

Let x = the size of the smallest angle.
Then 2x = the size of the medium-sized angle.
Then x + 2x + 90 = the sum of the angles of the triangle.
We know that the sum of the angles of the triangle equals 180°.

The equation is $\qquad$ x + 2x + 90 = 180
Add the x and the 2x $\qquad$ 3x + 90 = 180

Stop! I, your reader, object. We have never done this before. How do we know that x + 2x equals 3x?

You are right. This may be new, but I'd like to point out that you have been doing this *for years*.

One egg plus two eggs equals three eggs.
One cow plus two cows equals three cows.
One pizza plus two pizzas equals three pizzas.
One x plus two x's equals three x's.

Now, where were we? $\qquad$ 3x + 90 = 180
Subtract 90 from both sides $\qquad$ 3x = 90
Divide both sides by 3 $\qquad$ x = 30

The smallest angle is equal to 30°.

This is a very famous triangle. It's called the 30–60–90 triangle. In geometry we will prove that the shortest side is half as long as the longest side.

Complete Solutions and Answers | 315–317

315. Your kid sister has attached her pet 'gator to the ceiling with a spring.

The formula is F = kx, where F is the force on the spring (in this case, the weight of the alligator), where x is the distance the spring has stretched, and where k is a constant.

Suppose the alligator weighs 18 pounds and the spring has stretched 20 inches. What is the value of k?

$$F = kx \text{ becomes} \qquad 18 = k20$$

Put the number in front of the letter $\qquad 18 = 20k$

Divide both sides by 20 $\qquad \dfrac{18}{20} = k$

Simplify $\qquad \dfrac{9}{10} = k$

Or, if you like decimals, k = 0.9.

316. Solve each of these equations.

$$5 = \dfrac{x}{3}$$

Multiply both sides by 3 $\qquad 15 = x$

$$\dfrac{y}{9} = 6$$

Multiply both sides by 9 $\qquad y = 54$

$$\dfrac{w}{7} + 2 = 10$$

Subtract 2 from both sides $\qquad \dfrac{w}{7} = 8$

Multiply both sides by 7 $\qquad w = 56$

317. Will μ_s have changed?

The coefficient of friction, μ, depends only on which two surfaces are rubbing together. It doesn't depend on how hard the surfaces are pushed together. From the previous problem we know that μ_s was equal to $\dfrac{2}{3}$ when it was just the empty chest and the floor. Your standing on top of the chest will not affect μ_s.

| 318–319 | Complete Solutions and Answers |

318. I'm pushing a giant pineapple across a frozen lake. If the strength of gravity changes, will that affect how hard I have to push?

How hard I have to push depends on the friction between the pineapple and the ice. That friction depends on how hard the pineapple is pressing against the ice. It depends on the weight of the pineapple. If the strength of gravity is doubled, then the pineapple will press twice as hard against the ice, and the friction will double. I will have to push twice as hard.

319. You decide to tip the table and make it slide off.

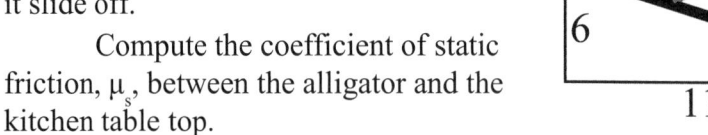

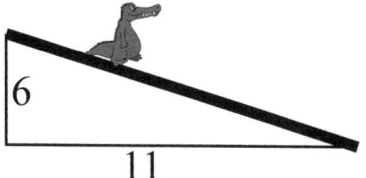

Compute the coefficient of static friction, μ_s, between the alligator and the kitchen table top.

❀ Express your answer as a fraction.

❀ Express your answer as a decimal rounded to the nearest one-hundredth.

In algebra we will call the 6 the rise and 11 the run. Kingie claimed that μ_s is equal to $\frac{\text{rise}}{\text{run}}$ We will show that this is true in the next chapter.

In the meantime, it's easy to use. $\mu_s = \frac{6}{11}$

Doing the arithmetic to change $\frac{6}{11}$ into a decimal takes a little more work.

$$\begin{array}{r} 0.545 \\ 11 \overline{)6.000} \\ -\underline{55} \\ 50 \\ -\underline{44} \\ 60 \\ -\underline{55} \end{array}$$

$0.545 \doteq 0.55$ ($\doteq$ means "equals after rounding")

Rounded to the nearest one-hundredth, $\mu_s = 0.55$

Complete Solutions and Answers | 333–340

333. On the floor is 4 pounds of socks and underwear. You lift them up 30 inches and put them on a shelf. How many foot-pounds of work was that?

Since we want to know foot-pounds, first we change 30 inches into feet.

$$\frac{30 \text{ inches}}{1} \times \frac{1 \text{ foot}}{12 \text{ inches}} = \frac{30}{12} \text{ feet} = \frac{5}{2} \text{ feet} = 2.5 \text{ feet}$$

Work = Force times distance
W = Fd becomes   W = (4 lbs.)(2.5 ft)
Doing the arithmetic   W = 10 ft-lbs

340. The density of the water became 0.59 oz./in^3. Will your 4-ounce CD float to the top of the water?

The volume of your CD with its case is 6.875 in^3. We computed that in the previous problem.

The buoyancy of any object = dv where density is the density of the liquid and v is the volume of the object in that liquid.

After she added the salt, the buoyancy would equal (0.59)(6.875), which equals 4.05625 ounces. Since the CD weighs 4 ounces, it will float to the surface; the upward buoyancy is greater than the downward weight.

Also floating to the top will be your dead fish. Fresh water aquarium fish don't do well in salt water.

344 Complete Solutions and Answers

344. After 24 observations, are you 100% certain that your parents care more for your kid sister than they do for you.? Are you using deductive reasoning?

You certainly are in a world of hurt. Almost anyone in your place would feel very bad. If I had parents like that, I would have tears in my eyes. To have a rotten kid sister is definitely not cool, but to have parents that don't care for you as much as for your sister, that can be devastating.

But the question is: **Is it true?**

Does being emotionally worked up about some issue change the truth of that issue? If you see protestors shouting and screaming and throwing things, does that affect whether their demands are righteous?

Item #1: Your kid sister's temper tantrums don't mean that her childish demands are right.

Item #2: I, your author, can state without much emotion that three plus four equals seven.

In the language of math: *Strong emotions does not always imply being right.*

But the question remains: **Is it true?** Do your parents care more about her than you?

The answer is . . . probably. With those 24 observations you might feel fairly certain that they care for her more than you. In a court of law, you could get the judge to rule in your favor.

But *my* question was *Are you 100% certain?*

We are going from observations, trials, and experiments and trying to arrive at a conclusion. This is inductive reasoning—not deductive reasoning.

You are using the same kind of reasoning that scientists use. And every decent scientist knows: *Science never proves anything with 100% certainty.*

Yeah. But, I your reader, knows that this is all just talk. Given those 24 observations of how your parents care for your kid sister more than they care for you . . . you have to know that that is true. There is no way that it couldn't be true.

122

Complete Solutions and Answers

I bet you are offering me a challenge. I like a challenge. When I wrote these first 29 chapters of this *Zillions* book, I also "knew" that your parents cared for her more than you. It was soooo obvious.

But . . . is that the only possibility?

First of all, how long do these 24 observations cover? We know she removed the chicken from the soup on New Year's Day (in Chapter 6). The other things might have occurred in only two or three days. We are not talking about three years of your parents showing you less care than your kid sister.

Is there any possible reason why you have been slighted for these two or three days? I can think of several reasons.

First possible reason: Many parents are human. (An understatement) Some weeks they might show extra concern for one of their kids, and in other weeks, they might direct their love toward another kid. Variation in human behavior is normal. Some days you like peanut butter and egg sandwiches, and other days you like tuna and radish sandwiches.

Second possible reason: Suppose on December 31st your parents learned from the doctor that your kid sister has a serious condition that can't be fixed—that she will be dead in a couple of weeks.

Your parents face a difficult decision as to what and when to tell your three-year-old sister. Your parents can't tell you what's happening. There is too much danger you might blurt out, "Hey. Did you know your funeral will be in two weeks?"

But, understandably, your mom and dad want to shower their love on the kid with the chubby hands for the next couple of weeks.

✓ They let her have all the chicken in the soup.
✓ They let her have any pet she wants, even it's an alligator.
✓ They let her watch all the three-year-old television programs she wants.
✓ They give her zillions of helium balloons because she likes them.
✓ They give her kisses and Sluice straws.
✓ They call her "little dearest daughter."

If you were in their place, you would probably do the same thing.

Those 24 observations might point to a different reality than your parents not caring for you.

| 352–360 | **Complete Solutions and Answers** |

352. In what region or regions will Hooke's law hold?

Hooke's law says that in the region up to the proportional limit, the amount that a spring is stretched, x, is proportional to the weight on the spring, F.

$F = kx$ k is called the spring constant.

356. How many foot-pounds of work did she do in raising that pizza 1½ feet? The weight of the pizza in water is 5.6 pounds.

The formula for work is W = Fd where F is the force (weight) and d is the distance.

W = Fd becomes   W = (5.6)(1½)   or   W = (5.6)(1.5)

Doing the arithmetic $W = 8\frac{2}{5}$ ft-lbs. or   W = 8.4 ft-lbs.

Here is the arithmetic.

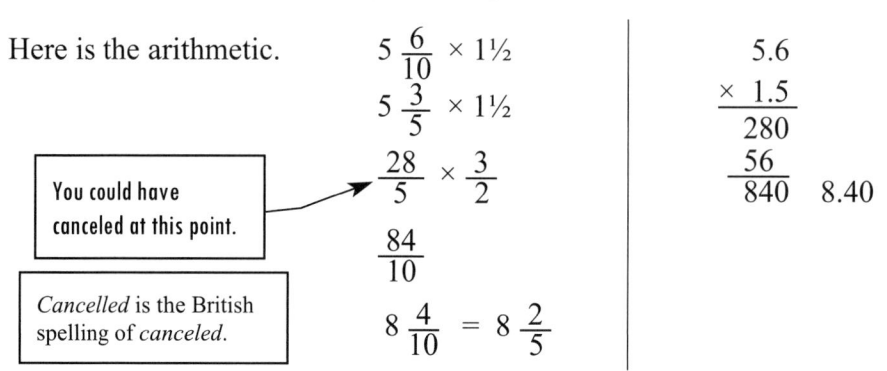

$5\frac{6}{10} \times 1\frac{1}{2}$

$5\frac{3}{5} \times 1\frac{1}{2}$

$\frac{28}{5} \times \frac{3}{2}$

You could have canceled at this point.

$\frac{84}{10}$

Cancelled is the British spelling of *canceled*.

$8\frac{4}{10} = 8\frac{2}{5}$

```
  5.6
× 1.5
  280
   56
  840   8.40
```

360. How much force will it take to get the glob and alligator moving?

$F = \mu_s N$ becomes   F = (0.7)(6 + 18)
Doing the arithmetic   F = (0.7)(24)
Doing more arithmetic   F = 16.8

It will take 16.8 pounds to get the glob and alligator moving.

Complete Solutions and Answers 364

364. One gram of fat is equal to 9 Calories. One gram of protein is equal to 4 Calories. Three pounds of protein can store as much Calories as how many pounds of fat? (1 pound ≈ 453 grams)

This sounds like a conversion factor problem to me.

We know that 1g of fat = 9 Calories.
The conversion factor will be either $\dfrac{1g\ fat}{9\ Calories}$ or $\dfrac{9\ Calories}{1g\ fat}$

We know that 1g of protein = 4 Calories.
The conversion factor will be either $\dfrac{1g\ protein}{4\ Calories}$ or $\dfrac{4\ Calories}{1g\ protein}$

We know that 1 pound = 453g.
The conversion factor will be either $\dfrac{1\ pound}{453g}$ or $\dfrac{453g}{1\ pound}$

Hey! I, your reader, think that that is a TON of conversion factors!

You are right. But once you have these conversion factors, the math will virtually write itself. Very little thought will be needed.

You are starting with 3 pounds of protein. That's what is given. (We will be looking for pounds of fat.)

What can you multiply $\dfrac{3\ lbs.\ of\ protein}{1}$ by?

There are six conversion factors that we have listed.
Only one of these six can cancel the lbs. So we use it.

$$\dfrac{3\ \cancel{lbs.}\ of\ protein}{1} \times \dfrac{453g}{1\ \cancel{pound}} = 1359g\ protein$$

Only one of these six will get rid of protein. So we use it.

$$\dfrac{1359g\ \cancel{protein}}{1} \times \dfrac{4\ Calories}{1g\ \cancel{protein}} = 5436\ Calories$$

I've now got Calories. I remember that I'm looking for pounds of fat.
Only one of these six conversion factors will convert Calories into fat. So we use it.

125

| 365 | Complete Solutions and Answers

$$\frac{5436 \text{ Calories}}{1} \times \frac{1\text{g fat}}{9 \text{ Calories}} = 604\text{g fat}$$

We have grams of fat and we are looking for pounds of fat.
Only one of these six conversion factors will convert grams to fat. So we use it.

$$\frac{604\text{g fat}}{1} \times \frac{1 \text{ pound}}{453\text{g}} = 1⅓ \text{ pounds of fat.}$$

So 1⅓ pounds of fat can store the same amount of energy that 3 pounds of muscle (protein) can store.

Two steps for doing conversion factor problems . . .

✓ Every equality (like 3 feet = one yard) offers two conversion factors:

$$\frac{3 \text{ feet}}{1 \text{ yard}} \quad \text{or} \quad \frac{1 \text{ yard}}{3 \text{ feet}}$$

✓ Start with the thing you are given. Keep in mind the goal. Use the conversion factor(s) that get rid of the units you don't want and that give you the units that you do want.

365. In physics, the letter c stands for the speed of what?

It is the speed of light. Not every letter in our alphabet—abcdefghijklmnopqrstuvwxyz—has a special meaning. But c does.

Not every letter in the Greek alphabet—αβγδεζηθικλμνξοπρστυφχψω—has a special meaning. But π does. π (pronounced pie) is the circumference of a circle divided by its diameter. It doesn't matter whether the circle is large or small. The answer is always the same. π is a little larger than 3.

Wait a minute! I, your reader, have a question. Didn't you make a mistake when you listed the letters in the Greek alphabet. I counted them. αβγδεζηθικλμνξοπρστυφχψω **has only 24 letters. You left out two letters.**

You are assuming something. It turns out that the Greek alphabet is easier to learn than abcdefghijklmnopqrstuvwxyz, which has 26 letters.

126

Complete Solutions and Answers 366–369

366. It took 8 steps (also known as 8 paces). How far is it from your oven to your bed? Use a conversion factor. (Your paces are 28".)

The conversion factor will be either be $\dfrac{1 \text{ pace}}{28"}$ or $\dfrac{28"}{1 \text{ pace}}$

$$\dfrac{8 \text{ paces}}{1} \times \dfrac{28"}{1 \text{ pace}} = 224 \text{ inches}$$

367.

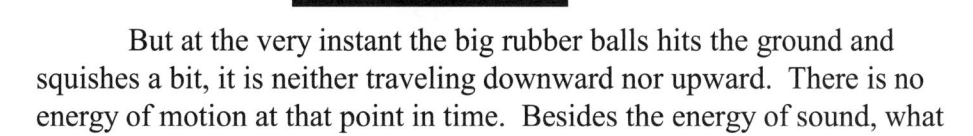

But at the very instant the big rubber balls hits the ground and squishes a bit, it is neither traveling downward nor upward. There is no energy of motion at that point in time. Besides the energy of sound, what has the motion energy been turned into?

It has turned into spring energy.

A moment later that spring energy will be turned back into motion energy as the ball bounces upward. You now know why big rubber balls bounce.

368. Your mom gave your kid sister a big vacuumy kiss on her cheek. Part of her cheek bulged out slightly. Where did the force come from that made her cheek bulge?

The normal atmospheric pressure that we experience at the bottom of this ocean of air (that's poetry!) is 14 psi (pounds per square inch).

Opposing that pressure is an outward pressure from our bodies. If we didn't have that outward pressure, we would be crushed.

Your mom, removed part of that atmospheric pressure. Your kid sister's outward pressure of 14 psi forced her cheek outward.

369. Simplify $\dfrac{893z}{893}$

If you take z and multiply it by 893 and then divide it by 893, you will get z back again. $\dfrac{893z}{893} = z$

| 383–393 | Complete Solutions and Answers |

383. You take the target off of the wall and drag it toward the garbage can. It weighs 8 pounds and takes 3 pounds of force pull it across your bedroom floor. Find μ_k.

$F = \mu_k N$ becomes $\qquad 3 = \mu_k 8$

Putting the number in front of the letter $\qquad 3 = 8\mu_k$

Dividing both sides by 8 $\qquad \dfrac{3}{8} = \mu_k$

Or, if you prefer decimals, $\mu_k = 0.375$

```
      0.375
   8) 3.000
     − 24
       60
     − 56
       40
     − 40
```

392. Solve $\qquad 7y = 48.3$

Divide both sides by 7 $\qquad y = 6.9$

The hardest part isn't the algebra. It is the arithmetic.

```
      6.9
   7) 48.3
      42
      63
      63
```

393. You pay the power company for the number of electrons you use. One ampere is defined as 6.24×10^{18} electrons per second.

Part A) If you run a 55 Ω light bulb on a 110 volt circuit, how many amperes will be flowing?

$V = IR$ becomes $\qquad 110 = I(55)$

Dividing both sides by 55 $\qquad 2 = I$

You will be using 2 amps.

Part B) How many electrons will you be using each second?

$$\dfrac{2 \text{ amps}}{1} \times \dfrac{6.24 \times 10^{18} \text{ electrons/sec}}{1 \text{ amp}} = 12.48 \times 10^{18} \text{ electrons/sec}$$

Complete Solutions and Answers | 395–398

395. You drag your 240-pound sack of carrots and find that it takes 80 pounds of force to pull it at a constant speed. What is the value of µ?

$$F = \mu N \quad \text{becomes} \quad 80 = \mu 240$$

In algebra we usually put the number before the letter

$$80 = 240\mu$$

Divide both sides by 240

$$\frac{80}{240} = \frac{240\mu}{240}$$

Simplify the right side

$$\frac{80}{240} = \mu$$

Simplify the left side

$$\frac{1}{3} = \mu$$

In this case it is easier to express µ as a fraction instead of a decimal. You may recall that ⅓ equals 0.333333333333333333333333333333333333333
3333333333333333333333333333333333333333333333333333333333333
3333333333333333333333333333333333333333333333333333333333333
3333333333333333333333333333333333333333333333333333333333333
3333333333333333333333333333333333333333333333333333333333333
3333333333333333333333333333333333333333333333333333333333333
3333333333333333333333333333333333333333333333333333333333333
3333333333333333333333333333333333333333333333333333333333333
3333333333333333333333333333333333333333333333333333333333333
33333333333333333333333333333333 etc.

398. If the atmosphere of earth were pure helium instead of a mixture of oxygen and nitrogen and we repeated the glass tube experiment, what would be the result?

When the glass tube experiment is done in air, the water falls to 34 feet above the surface of the lake because the weight of the air can push the column of water 34 feet upward.

Helium is much less dense than air. The column of water will be lower.

404–406 Complete Solutions and Answers

404. Some people have a lot of ducks in their living room. Some people have few ducks in their living room. About seven billion people (7,000,000,000) have zero ducks in their living room. If we let x equal the number of ducks in a person's living room, is x a discrete or a continuous variable?

For some people, x = 12. For most people, x = 0. If x were a continuous variable, then x could take on values such as x = 7.39. This is not possible. x is a discrete variable.

405. She drags the 24-pound tank across your bedroom floor at 4 mph. It takes 8 pounds of force to push the tank. Find either the coefficient of static friction or the coefficient of kinetic friction or find both.

We can find μ_k (the coefficient of kinetic friction) since $F = \mu_k N$.

$F = \mu_k N$ becomes	$8 = \mu_k 24$
Put the number in front of the letter	$8 = 24\mu_k$
Divide both sides by 24	$\frac{1}{3} = \mu_k$

We can't find μ_s because we don't know how hard she had to push to get that tank started. All we know is that it takes more push to get something started than it does to keep it moving. We know $\mu_s > \frac{1}{3}$.

406. How hard will your sister have to push to get you and the chest moving?

We know $F = \mu_s N$. We know that $\mu_s = \frac{2}{3}$. We know that N = 171 pounds.

$F = \mu_s N$ becomes	$F = (\frac{2}{3})(171)$
Doing the arithmetic	$F = 114$ pounds

$$\frac{2}{3} \times \frac{171}{1} = \frac{342}{3} \qquad 3\overline{)171}^{\,114}$$

Complete Solutions and Answers | 420–422

420. All five of your senses tell you that it's an alligator. Can you be 100% certain?

Guess what. You were wrong.
It was just Fred wearing an alligator costume.

421. Hint #1: You start this kind of problem by letting x equal the thing you are trying to find out. In this case, let x = the weight of each coat of paint.
Hint #2: The three coats of paint will weigh 3x.
Hint #3: After applying the paint and sticking her doll inside, the weight will have increased by 3x + 8.
Hint #4: We are told that the weight increased by 17.
Hint #5: 3x + 8 = 17

$$3x + 8 = 17$$
Subtract 8 from both sides $\qquad 3x = 9$
Divide both sides by 3 $\qquad x = 3$

Each coat of pink paint weighed 3 pounds.
  That's a lot of paint!

422.

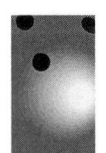

They weigh the same. They are given the same push. The only question is which one of them has more friction.
  What Fred pulled on his tiny safe across his desktop he discovered that it didn't make any difference whether the safe was upright or on its side. The amount of friction does not depend on how much contact (how much area) they have.
✓ Friction will depend on the surfaces. (It's easier to slide on ice than on carpet.)
✓ Friction will depend on how hard the objects are pressed together. (It's easier to slide fork across a table than to slide your sister across the table.)
✓ Friction does *not* depend on the area of contact. (It's just as easy to slide your sister on her side as it is to slide her on her back, unless she's ticklish.)

| 424–426 | Complete Solutions and Answers |

424.

In this schematic . . .

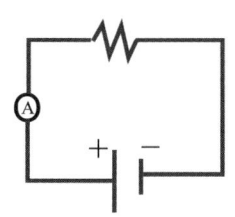

Question A) If I = 7 A and R = 8 Ω, what is the voltage of the battery?

    V = IR becomes   V = (7)(8)
    Doing the arithmetic   V = 56 volts

Question B) If R = 16Ω and V = 8 V, what does the ammeter read?

    V = IR becomes   8 = I(16)
    Putting the number in front of the letter   8 = 16I
    Dividing both sides by 16   0.5 = I

The ammeter will read 0.5 amps (or ½ amps if you like fractions)

Question C) If V = 110 volts and I = 37 A, what will the ammeter read?

    This question will test whether you are awake. The ammeter will read 37 A.

426. When she was underwater (or rather underoil), her buoyancy was 80 pounds. Since she doesn't weigh 80 pounds, she floated to the surface. What is the volume of your kid sister? (Recall, the density of the oil is 50 pounds/cubic foot.)

    buoyancy equals the density of the liquid times the volume of the liquid that is displaced.

    b = dv becomes   80 = 50v
    Divide both sides by 50   1.6 = v

The volume of your kid sister is 1.6 cubic feet.

The arithmetic:
$$50 \overline{)80.0} = 1.6$$
            50
           300
           300

Complete Solutions and Answers — 428–445

428. Your brain thinks. Where does it get its energy from?

If you want the whole story, we have to start with the nuclear reactions in the sun that turn nuclear energy into light.

Then the light travels about eight minutes and gets to the earth and hits green plants. That light is transformed into sugars, starches and oils in the plant. (That's called photosynthesis.) The energy in those sugars, starches, and oils is chemical energy.

You either eat the plants or eat the animals that ate the plants. In either case through a bunch of chemical reactions in your mouth, tummy, and small intestines, the food is turned into amino acids and glucose in your blood. Also mixed into your blood is oxygen from your lungs. (That's why it's important to breathe every once in a while.)

The whole mess is delivered in your blood to your brain cells (Thank you heart!). More chemical reactions inside the brain cells and your thoughts happen.

Be thankful for the sun, for the plants and photosynthesis, for animals that eat the plants, for pepperoni pizza, for having enough money to buy the pizza, for your tummy, and for your heart.

445. With her little chubby feet all coated with oil, she could pretend she was skating on the concrete near the pool. She ran and then slid. She shouted, "Wheeeeee!" and came to a stop. μ_k is very low between oily feet and concrete. The energy of motion was converted into what other form of energy?

When two things rub together heat is generated. In fact, by the time she came to a stop, the undersides of her feet was noticeably warmer. You knew this because she asked you to feel her feet.

| 449–452 | Complete Solutions and Answers

449. Suppose you could tip the oven rack and found that the pizza would start to slide when the rack was tipped at 27°. Is that enough information for you to determine μ_k?

No. Kingie's special formula, $\frac{\text{rise}}{\text{run}}$, was good for finding μ_s, which is the coefficient of static friction. In trig we will learn that tan 27° equals rise/run.

What that 27° will give you when you have a scientific calculator with a **tan** button is the value of μ_s (and not μ_k).

450. Your pizza has a volume of 0.2 cubic feet. What is the upward force (the buoyancy) of your pizza?

The formula buoyancy = dv becomes buoyancy = (62)(0.2)
Doing the arithmetic buoyancy = 12.4 lbs.

451. In 2 hours Fred can run 9 miles. Plot on a graph the point (2, 9).

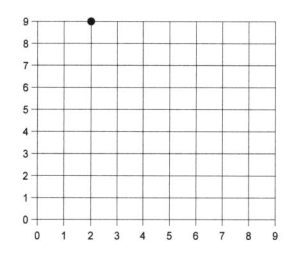

452. How far is 224" in feet? How far is that in feet and inches?

$$\frac{224 \text{ inches}}{1} \times \frac{1 \text{ foot}}{12 \text{ inches}} = \frac{224}{12} \text{ feet}$$

If you like to work in fractions $\frac{224}{12}$ feet becomes 18⅔ feet.

If you like to work in decimals $\frac{224}{12}$ feet becomes $12\overline{)224}$ = $18.66\overline{6}$

To convert 224" into feet and inches we divide and use a remainder.

$12\overline{)224}$ = 18 R8   224" = 18' 8"

134

Complete Solutions and Answers | 453–455

453. It takes 10 pounds to move the balloon and alligator at 8 mph across the floor. How much force is needed to move them at 16 mph?

The constant of kinetic (moving) friction, μ_k, doesn't depend on what speed you are pulling/pushing the object. It only depends on the two surfaces that are in contact. It will take 10 pounds to move the glob and alligator at any particular constant speed.

If you are pulling them at 8 mph, it will take a bit of extra force to get them moving at 16 mph, but once they are at the new speed, then you can relax a bit and keep them at the new speed with 10 pounds of force.

454. You drag 60 pounds down the street toward the library. It takes 20 pounds of force to drag it at 4 mph.

As you are dragging it, you reach into your backpack and take out 3 pounds of beef jerky and eat it. How much force will it now take to drag 57 pounds at 4 mph?

A two-step problem.
First, we find the coefficient of sliding (kinetic) friction, μ_k.

$F = \mu_k N$ becomes $\qquad 20 = \mu_k(60)$

Putting the number in front of the letter $\qquad 20 = 60\mu_k$

Dividing both sides by 60 $\qquad \dfrac{20}{60} = \mu_k$

Doing the arithmetic $\qquad \dfrac{1}{3} = \mu_k$

Second step, we find the force need to drag 57 pounds.

$F = \mu_k N$ becomes $\qquad F = \dfrac{1}{3} \times \dfrac{57}{1}$

Doing the arithmetic $\qquad F = 19$ pounds

455. He and his board weigh 140 pounds. Mu (μ) between the board and the sand is 0.6. How hard will you have to push?

$F = \mu W$ becomes $F = (0.6)(140)$
$$\begin{array}{r} 140 \\ \times\ 0.6 \\ \hline 840 \end{array}$$

84.0 pounds are required.

456–458 Complete Solutions and Answers

456. Which is the larger number, 3 or 9?

You kid sister might say that 3 is larger (since she's three years old.) But she would be wrong. Nine is a larger *number* than three.

What your kid sister was confusing was the concepts of *number* and *numeral*. A numeral is the physical thing you write down. Certainly, 3 is a larger numeral than 9.

457. It takes 10.8 pounds to slide your chest of drawers at 4 mph across the carpet in your bedroom. Your chest of drawers weighs 18 pounds. What is the coefficient of friction, μ?

$F = \mu N$ becomes $10.8 = \mu(18)$
Putting the number in front of the letter $10.8 = 18\mu$
Dividing both sides by 18 $\dfrac{10.8}{18} = \mu$

$$\begin{array}{r} 0.6 \\ 18\overline{)10.8} \\ -\underline{108} \\ 0 \end{array}$$

Simplifying $0.6 = \mu$

458. It is often inconvenient to erect a 35-foot tall glass tube in your bedroom, fill it with water, and notice that the water falls to 34 feet. That indicates a pressure of 14.7 lbs./in². The density of mercury is about 0.49 pounds per cubic inch. It's heavy! How many inches of mercury is needed to create a pressure of 14.7 lbs./in²?

The formula relating pressure, density, and height is $p = dh$.

$p = dh$ becomes $14.7 \text{ lbs./in}^2 = (0.49 \text{ lbs./in}^3)h$
Divide both sides by 0.49 $30 \text{ inches} = h$

The column of mercury would be 30 inches tall. That is easy to fit in your bedroom.

$0.49\overline{)14.7}$ becomes $\begin{array}{r} 30. \\ 49\overline{)1470.} \\ \underline{147} \\ 0 \end{array}$

Complete Solutions and Answers | 459–461

459. #1 reads 15 A.
#2 = 10 Ω.
#4 = 15 Ω.
What will ammeter #3 read?

It doesn't matter how many ohms #2 and #4 have. If 15 amperes are going through #1, then 15 amperes must be going through #3.

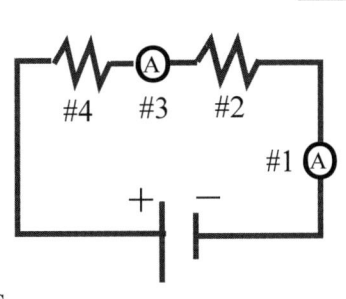

460. You did 5 ft-lbs. of work in lifting that book 30 inches. How heavy is that book?

From the previous problem, we know 30 inches = 2½ feet.

W = Fd becomes   5 ft-lbs. = F(2½ feet)

Dividing both sides by 2½   5 ÷ 2½ = F

Doing the arithmetic   2 lbs. = F

Here is the arithmetic: 5 ÷ 2½

$$\frac{5}{1} \div \frac{5}{2}$$
$$\frac{5}{1} \times \frac{2}{5}$$
$$2$$

461. In this diagram, Ⓐ is an ammeter, which measures how many electrons are flowing past that point in the circuit.

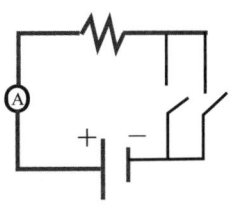

Which is true?
  A) There are zillions of electrons flowing through the ammeter.
  B) There is no way to tell how many electrons are flowing through the ammeter.
  C) There are zero electrons flowing through the ammeter.

Both of those switches are open. There is no way that electrons can run around in that circuit. The answer is still C).

137

485–487 Complete Solutions and Answers

485. You are talking to your friend on Mars trying to explain over the telephone exactly how much a kilogram is. The only way you can communicate is by voice. You can't mail him a 1-kilogram bar.

Why wouldn't this approach work? *Find something that weighs 2.205 pounds. That's a kilogram.*

That would work if you were talking with someone in India or in Wyoming or anywhere where the gravity is the same as yours. But Mars is a smaller planet than earth. The gravity is weaker.

A second objection is your friend on Mars would not necessarily know what a pound was.

> He would know what a dozen was since you could count together on the phone.
> He would know what c was because that speed is the same on Mars as it is on earth.
> He would know what the color orange was since it is defined to be light with wavelength 0.0000005970 to 0.0000006220 meters.

486. If the area of a rectangle is 15 and the width is 10, what is its length?

$A = \ell w$ becomes $15 = \ell(10)$

or $15 = 10\ell$ since in algebra we often put the number in front of the letter.

Divide both sides by 10 and we get $\frac{15}{10} = \ell$

Simplifying $\frac{15}{10} = 1.5$ or $\frac{15}{10} = \frac{3}{2} = 1\frac{1}{2}$

Some people like to work in decimals and some people like fractions.

487. What does $0.08v - 0.01v$ equal?

Recall: 3 eggs + 5 eggs equals 8 eggs
          6 moose − 2 moose = 4 moose
          70 pies − 2 pies = 68 pies

So $0.08v - 0.01v$ equals $0.07v$. ☺

138

Complete Solutions and Answers | 488–489

488. If Fred's swim mask were rectangular and the width was 2½ inches, what would be the length? Do your work in fractions for this problem. Recall, the area of the mask is 10 in².

$A_{rectangle} = \ell w$ becomes $10 = \ell(2½)$
Divide both sides by 2½ $10 \div 2½ = \ell$
Doing the arithmetic $4 = \ell$

$$\frac{10}{1} \div 2½$$

$$\frac{10}{1} \div \frac{5}{2}$$

$$\frac{10}{1} \times \frac{2}{5} = \frac{\overset{2}{\cancel{10}}}{1} \times \frac{2}{\cancel{5}_1} = 4$$

The length of the mask is 4 inches.

489. As you drag it across the floor, some of the boogers that are on top of the target (not in contact with the floor) drop off.

You are now pulling a load of 7 pounds instead of 8. How many pounds of force will it take to keep it moving at 2 mph?

In the previous problem we found that $\mu_k = \frac{3}{8}$

Since the two surfaces have not changed, μ_k has not changed. (It would have changed if the target had been dragged face down with the boogers in contact with the floor. Then as they dropped off, the coefficient of kinetic friction would have changed because one of the surfaces would have changed. Everyone knows that paper with boogers slides more easily than just plain paper.)

$F = \mu_k N$ becomes $F = (\frac{3}{8})7$

Doing the arithmetic $F = \frac{21}{8} = 2\frac{5}{8}$ pounds

In *Life of Fred: Decimals and Percents* you memorized the Nine Conversions. One of them is $\frac{5}{8} = 62.5\%$

So in decimals, the forced needed to drag the 7 pounds is 2.625 pounds.

139

| 490 | Complete Solutions and Answers |

490.

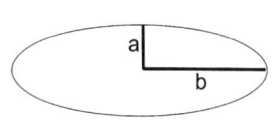

You point out that circles are just ellipses in which the a and b distances are the same. If you know that $A_{ellipse} = \pi ab$, then what must be the area of a circle?

If a and b are both equal to r, then $A_{ellipse} = \pi ab$ becomes $A_{circle} = \pi rr$

In algebra, we are going to abbreviate rr by r^2. $A_{circle} = \pi r^2$

Hey! I, your reader, think that's kind of silly. Are you really saving much space by writing r^2 instead of rr?

I agree. But when in algebra you have rrrrrrrrrrrr, it's much nicer to write r^{12}. ☺

But when you are ever going to need rrrrrrrrrrr? That will never come up.

Oh? You live in a universe—all the stuff of plants, anteaters, rivers, the moon, the sun, the stars.

Of course. But what has that to do with rrrrrrrrrr?

Easy. Scientists estimate that there are 10,000,000,000,000,000,000,000,000,000,000,000,000,000,000,000,000,000,000,000,000,000,000,000,000,000,000,000 particles in the observable universe. I think it's much easier to write 10^{79}.*

---

* 10^{79} means
10×10×10×10×10×10×10×10×10×10×10×10×10×10×10×10×10×10×10×
10×10×10×10×10×10×10×10×10×10×10×10×10×10×10×10×10×10×
10×10×10×10×10×10×10×10×10×10×10×10×10×10×10×10×10×10×
10×10×10×10×10×10×10×10×10×10×10×10×10×10×10×10×10×10×
10×10×10.

Complete Solutions and Answers | 491–494

491. Joules (pronounced JEWELS) is a measure of work in the metric system. What is the corresponding unit in the Imperial system?

    It is the unit that we have been working with since Chapter 16. The unit is foot-pounds.

492. You are given 36 = 5y. Find the value of y.

You start with	$36 = 5y$
Divide both sides by 5	$\frac{36}{5} = \frac{5y}{5}$
Simplify the right side	$\frac{36}{5} = y$
Simplify the left side	$7\frac{1}{5} = y$

$$5\overline{)36} \quad \begin{array}{l} 7 \text{ R1} \\ -35 \\ \hline 1 \end{array} \quad \text{or } 7\frac{1}{5}$$

493. You have just determined the buoyancy of your pizza. If your pizza weighs 18 pounds. How hard will it be pressing against your mom's legs?

    The pizza's weight is 18 pounds. That's pressing ↓.

The buoyancy of 12.4 pounds is pushing ↑.

    Subtracting   18.0 downward
                          12.4 upward
                          5.6 pounds net downward pressure

494. Consider your average ice-skating duck. If you are pushing him, does it take more effort if he has one foot in the air or if he has both feet on the ice?

    The force necessary to push a duck on the ice is independent of the area of contact between the duck and the ice.

| 495–497 | Complete Solutions and Answers

495. In raising that pizza she converted what form of energy into what other form of energy? The nine forms of energy are chemical, electrical, heat, height, light, motion, nuclear, sound, and spring.

Her muscles use chemical energy. (It's a known fact that if you didn't eat for six years, you wouldn't be able to move your muscles.)

She converted that chemical energy into motion energy into the energy of height.

496. Your big dead fish weighs three times as much as your little dead fish. Together they weigh 13 ounces. How much does your little dead fish weigh?

The first step in any word problem is to let x = the thing you are trying to find out. Let x = the weight of the little dead fish.

Then 3x = the weight of the big dead fish.

Together they weigh 13 ounces.

At this point you can write the equation. It is *after* you have written the "Let x =" and *after* you have written "Then . . ." that you should write the equation. I know. That means that you have to write English stuff down on your paper. Get in the habit now. Later, in algebra, when things get a bit more complicated, some students attempt to jump from the word problem directly to the equation and they fall into a hole.

The equation is x + 3x = 13

Combine x and 3x $4x = 13$

Divide both sides by 4 $x = \dfrac{13}{4}$

Do the arithmetic $x = 3\dfrac{1}{4}$ or   x = 3.25 ounces.

497. You lift that 5-ounce bar of silver 18 inches upward so that you can get a good look at it. Does that involve the same amount of work as when you lifted your 5-ounce bar of chocolate 18 inches upward to take a sniff of it?

Work is equal to force times distance. Both the chocolate and the silver were lifted the same distance. But the force changed since 5 ounces (avoirdupois) is less than 5 ounces (troy) so the work involved in lifting the silver was more than the work needed to lift the chocolate.

Complete Solutions and Answers | 498–499

498. Now suppose the alligator eats your kid sister's sticker collection. He now weighs 27 pounds. How far will the spring stretch?

The formula for stretching a spring is $F = kx$.
We know that k is 0.9 from the previous problem.
We know that F is now 27.

$F = kx$ becomes $\qquad 27 = 0.9x$

Divide both sides by 0.9 $\qquad 30 = x \qquad 0.9 \overline{)27}$

$$9\overline{)270.}^{\;30}$$

The spring stretches 30 inches.

If you did your work in fractions, it would look like this . . .

$$27 = \frac{9}{10}x$$

Divide both sides by $\frac{9}{10}$ $\qquad 30 = x \qquad 27 \div \frac{9}{10}$

$$\frac{27}{1} \times \frac{10}{9}$$

$$\frac{\overset{3}{\cancel{27}}}{1} \times \frac{10}{\underset{1}{\cancel{9}}}$$

$$30$$

499. Suppose we have a triangle in which all three angles are equal. What is the size of each of those angles?

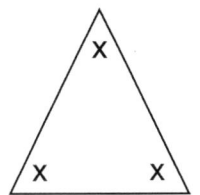

Let x = the size of one of those angles.
Then $x + x + x$ = the sum of all three angles.
We know that the sum of all three angles is 180°.

The equation is $\qquad x + x + x = 180$
$\qquad\qquad\qquad\qquad\qquad 3x = 180$
$\qquad\qquad\qquad\qquad\qquad\; x = 60$

Each of those angles is equal to 60°.

In geometry we will call that triangle an equiangular triangle. It is also called an equilateral triangle. (equi = equal, lateral = sides)

500–502   Complete Solutions and Answers

500. Her 32-ounce pillow floated. The density of that salt water is 0.59 oz./in³. What is the volume of the part of the pillow that is underwater? Round your answer to the nearest cubic inch.

buoyancy = dv   where d is the density of the salt water and v is the volume that is submerged. We are looking for v.

Since the pillow is floating, the buoyancy is equal to the weight of the pillow. (If the buoyancy were less than the weight of the pillow, the pillow would sink farther into the water. If the buoyancy were greater than the weight of the pillow, the pillow would rise.)

$$\text{buoyancy} = d_{\text{salt water}} \, v_{\text{of the submerged part of the pillow}}$$

becomes   weight of the pillow = $(0.59)(v_{\text{of the submerged part of the pillow}})$

becomes   32 oz. = $(0.59 \text{ oz./in}^3)(v_{\text{of the submerged part of the pillow}})$

Dividing both sides by 0.59 $54.237 = v_{\text{of the submerged part of the pillow}}$

Rounding to the nearest ounce $54 \text{ in}^3 = v_{\text{of the submerged part of the pillow}}$

501. (continuing the previous problem) What is the total bill from the carpenter?

Labor	$ 1,800.00
Materials	$ 186.92
7% sales tax on materials	$ 13.08
TOTAL	$2,000.00

502. If I smash that pineapple down, its weight will not change, but more of it will be in contact with the ice. How will that affect how hard I have to push?

The area of contact has no effect on the friction, and, hence, no effect on how hard I have to push. In the first chapter Fred experimented and found that a safe sitting upright and a safe on its side required the same effort to pull it.

Complete Solutions and Answers | 503–504

503. Physicists (at least today) think that all electrons in the universe are all alike. The happy little electron ☺ who leaves the negative end of the battery is virtually identical to the happy little electron ☺ who has made the trip through the light bulb and is arriving back at the positive end of the battery. Each morning when you wake up, you are different than the previous morning. You are not a happy little electron ☺. Name two ways that you are different.

    First of all, you are older. Second, when you wake up on Wednesday, you have memories of Tuesday. When you woke up on Tuesday, you didn't have memories of the day you were about to experience.

    On Wednesday you have some new skills, some new scratches, some new hopes that you didn't have on Tuesday.

    The truth is that electrons are not very happy at all. They have no chance to grow. Everything is frozen in time for them. ☹

504. How about now?

A) There are zillions of electrons flowing through the ammeter.
B) There is no way to tell how many electrons are flowing through the ammeter.
C) There are zero electrons flowing through the ammeter.

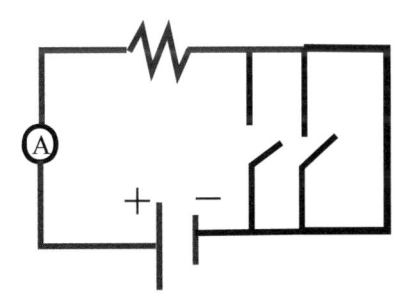

    Now there is a path for the electrons to run in. A) There will be zillions of electrons flowing through the ammeter.

    Let me take my gray pencil and trace the path.

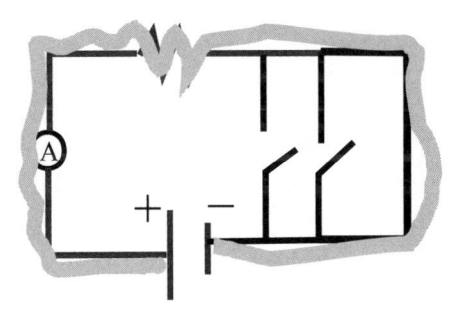

145

| 505–507 | Complete Solutions and Answers |

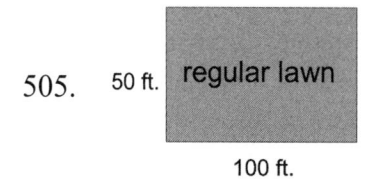

505.   50 ft. regular lawn / 50 ft. (ellipse)
100 ft.   /   100 ft.

What is the area of a "regular lawn"?
What is the area of your elliptical lawn?
(Use $\pi = 3$ for this problem.)

$A_{rectangle} = \ell w = (100)(50) = 5{,}000 \text{ ft}^2$
$A_{ellipse} = \pi ab = (3)(25)(50) = 3{,}750 \text{ ft}^2$

506. The hole she drilled is in the shape of an ellipse, with the width being 1" and the height being 0.4". What is the area of that ellipse?

If the major axis is 1" and the minor axis is 0.4", then $a = 0.2$" and $b = 0.5$".

(The major axis of an ellipse is the longer dimension. The minor axis is the shorter dimension.)

The area formula is $A_{ellipse} = \pi ab$ becomes $A = \pi(0.2)(0.5)$
Doing the arithmetic $A = 0.1\pi \text{ in}^2$

$0.1\pi \text{ in}^2$ is the *exact answer*.

If we were told to use $\pi \approx 3$, then the answer would be 0.3 in^2.
If we were told to use $\pi \approx 3.1415926535897932384626433832795$, then the answer would be $0.31415926535897932384626433832795 \text{ in}^2$.

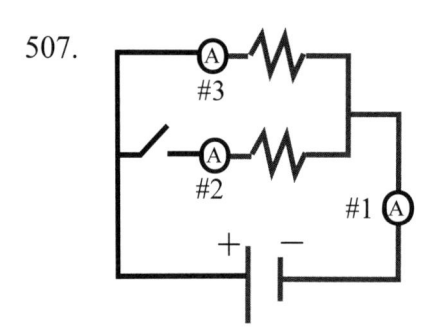

507.

#3 reads 17 amperes.

What do #1 and #2 read?

No electrons are passing through #2 since the switch is open. #2 will read 0 amperes.

Every electron that passes through #3 will pass through #1. #1 will read 17 amperes.

Complete Solutions and Answers — 534

534. Looking at the graph you drew, estimate how many waffles you could make in 8 minutes.

First, draw a straight line through the two points that you have already graphed.

Second, we want to see how many waffles correspond to 8 minutes. Draw a line upward (not *upwards*) from 8 minutes.

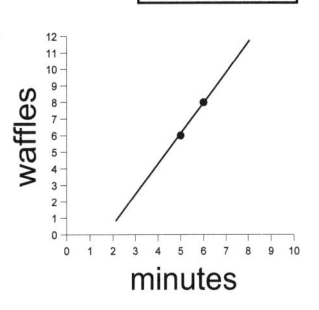

What number of waffles corresponds with 8 minutes? I draw one more line.

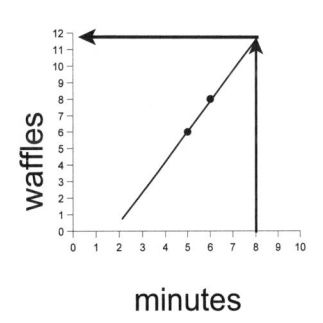

My best guess is that 8 minutes will correspond with 12 waffles.

Discussion: Many things in life go in a straight line. A straight line is often the best guess.

But not always.

Your brother can paint a fence in one hour. He can paint two fences in two hours. But can he paint 10 fences in 10 hours? No. He will get tuckered out.

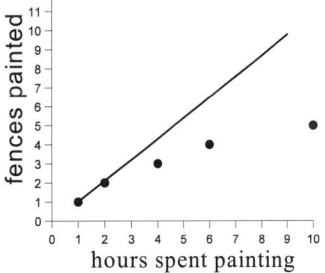

147

| 535–536 | **Complete Solutions and Answers** |

535. You straighten the stack up. Does the volume decrease, increase, or stay the same?

Cavalieri's principle* says that the volume of a stack of poker chips equals the area of any one of the poker chips times the height of the stack. It doesn't depend on whether the chips are all lined up vertically or not. Straightening the stack of pizzas will not change its volume.

536. Let's suppose that you have a really large number for the spring constant. You have a spring where F = 1,000,000x where F is measured in pounds and x is measured in inches.

If you hang a ten-pound weight on this spring, how much will the spring be stretched?

$$F = kx \text{ becomes} \qquad 10 = 1,000,000x$$

Divide both sides by 1,000,000 $\qquad \dfrac{10}{1,000,000} = x$

Canceling $\qquad \dfrac{1}{100,000} = x$

or in decimals x equals 0.00001 inches.

That is what you might call a very stiff spring. A ton (2,000 pounds) would stretch it two-thousands of a inch.

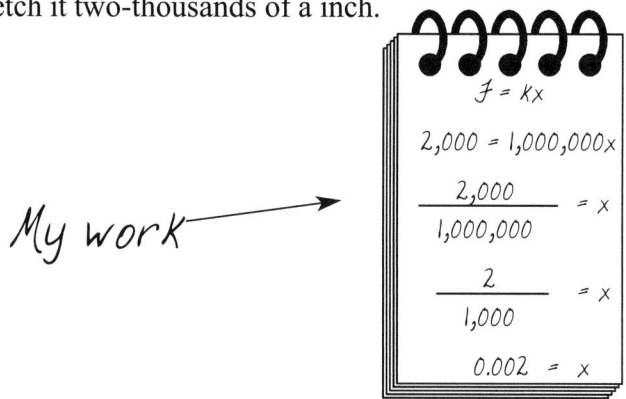

My work ⟶

$F = kx$
$2,000 = 1,000,000x$
$\dfrac{2,000}{1,000,000} = x$
$\dfrac{2}{1,000} = x$
$0.002 = x$

---

* When Cavalieri was in high school, his *principal* gave him an award because of his invention of the Cavalieri's *principle*. The *principal* said that the *principal* reason he gave him the award was that it was one of his *principles* to award good inventions.

Complete Solutions and Answers | 544–564

544. In the metric system 6 m stands for six meters. If you wrote six meatballs, you were N.C. (not close)

555. It takes 10.8 pounds to slide your chest of drawers at 4 mph across the carpet in your bedroom. If I slide your chest of drawers at 8 mph, will the force needed be 5.4, 10.8, or 21.6 pounds?

The force needed to slide an object at a constant speed doesn't depend on what that speed is. It will take 10.8 pounds *at any constant speed*.

560. Your pet duck is growing up. It's time to get him some shoes.

The shoe sizes for ducks are 5, 5½, 6, 6½, 7, 7½, 8, 8½, 9, 9½, 10, 10½, and 11. Is this a discrete or continuous variable?

It is a discrete variable. Sizes 7 and 7½ are available, but 7⅓ is not. This is the test for continuous variables. If two different numbers are possible, then every number between them must be possible.

564. There is energy stored in the oak that is released as heat when it burns. Which of the nine forms of energy is used to store energy in the oak? *motion/heat/light/sound/electrical/height/nuclear/spring/chemical*

It's not motion. You don't see the oak log vibrating or moving.
It's not heat. Before it is ignited, it is at room temperature.
It's not light. A log of wood makes a very poor lamp.
It's not sound. Put your ear up to it and you don't hear ♬♪.
It's not electrical. No one ever got electrocuted holding onto a log.
It's not height. The log isn't sitting up on some perch.
It's not nuclear. Materials like uranium and radium have atoms that can be split to release nuclear energy.
It's isn't a spring. You can't compress a log and expect it to pop back when you let it go.

It is chemical. Sunlight is converted by photosynthesis into the carbon compounds in wood. They hold the energy until you ignite the log.

| 570–572 | **Complete Solutions and Answers**

570. You wake up in the morning. You are in your bed. You decide that you are only going to do things where you are 100% certain of their outcome.

    i) Do you get out of bed and walk across the room?

    ii) Do you pour some cereal into your bowl?

    iii) Do you ask Pat to marry you?

    No, no, and no.

i) You don't know for certain that your kid sister hasn't released a dozen alligators in your bedroom, which would eat you alive before you got across the room. There is only a 0.00001% chance she's done that. The alternative would be to stay in bed for the rest of your life.

ii) Are you 100% sure that she hasn't put tiny alligators into your cereal box. If you poured them into your bowl, those 'gators might eat your spoon. Of course, there's only a 0.0000000000038% chance she's done that. The alternative would be to starve to death.

iii) What if Pat were really an alligator dressed up in a people costume? You would be marrying an alligator. Yuck! Of course, there is only a 0.000000000000000000000000000000000007% chance that would happen. The alternative is to tell Pat that you don't want to marry because you might be marrying an alligator. Then Pat would certainly not want to marry you because Pat would know that you are nuts.

571. 1 kilometer (km) = 1,000 meters (m) = 10^3 m.

One million dollars = $1,000,000
= $10 \times 10 \times 10 \times 10 \times 10 \times 10$ = $\$10^6$

572. From the previous problem we know how hard your kid sister will have to push to get you and the chest moving. F = 114 pounds.

μ_s between her shoes and the carpet in your bedroom is 0.3.

    $F = \mu_s N$ becomes   114 = 0.3N

    Divide both sides by 0.3   380 = N

That will need a lot of rocks in her pockets. She and her gown will have to weigh 380 pounds.

Complete Solutions and Answers | 573

573. Does the buoyancy of the pizza in the bath water depend on how deep the pizza is below the surface?

The buoyancy of an object is given by the formula buoyancy = dv where d is the density of the water (or other fluid) and v is the volume of the object that is submerged.

Does the density of water, d, change from the top of the water to the bottom of the tub? Short answer: No.

Does the volume of the pizza, v, change from the top of the water to the bottom of the tub? Short answer: No.

So if d and v aren't changing, then buoyancy, which is equal to dv, doesn't change.

Technical discussion. You can compress air. When you stick air in an auto tire you are compressing it. If you sit on a balloon, you can squeeze it down a bit (unless you pop the balloon). You can change the density of air. When people climb to the top of very tall mountains, the air is very thin (less dense) than down at sea level.

Can you compress water? If your kid sister asks you, you should tell her no. If your mom, your dad, your uncle, or your teacher asks you, you should tell them water is compressible.

There is an easy proof that water is compressible. Just dive underwater in a pool or in your bathtub and knock against the side. You can hear it. Sound travels through water. Sound travels through air or through water in sound waves. These waves are waves of compression.

How compressible is water? very little

In metric: If you were to swim 4 km (kilometers) down in a deep part of the ocean where the pressure is 40 MPa (40 million Pascals), water would be compressed to 98.2% of its original volume.

Translation from metric to the Imperial system (used in the U.S.A.): If you were to swim 2.5 miles down in a deep part of the ocean where the pressure is 5,802 psi (pounds per square inch), water would be compressed to 98.2% of its original volume.

Of course, you are not water. You have, for example, air in your lungs. You are more compressible than water. At 5,802 psi, you would be significantly compressed. Translation: You would be dead.

| 584–589 | Complete Solutions and Answers |

584. Suppose you want to push him at 9 mph instead of 3 mph. How hard would you have to push?

We just computed that it would take 84 pounds to push him at 3 mph. In Chapter 2 Fred learned that friction is independent of speed. It will take 84 pounds to maintain the scared surfer at any constant speed on the sand.

588. One ampere represents a flow of 6.24×10^{18} electrons per second. How many electrons per minute is that? (Use a conversion factor.)

The conversion factor will be either $\dfrac{1 \text{ minutes}}{60 \text{ seconds}}$ or it will be $\dfrac{60 \text{ seconds}}{1 \text{ minute}}$

$$\dfrac{6.24 \times 10^{18} \text{ electrons}}{1 \text{ second}} \times \dfrac{60 \text{ seconds}}{1 \text{ minute}}$$

$$= \dfrac{6.24 \times 10^{18} \text{ electrons}}{1 \text{ second}} \times \dfrac{60 \text{ seconds}}{1 \text{ minute}}$$

$$= 6.24 \times 10^{18} \times 60 \text{ electrons}$$
$$= 374.4 \times 10^{18} \text{ electrons}$$

Some people might write 374.4×10^{18} as $3.744 \times 10^2 \times 10^{18}$ or as 3.744×10^{20} or as $374{,}400{,}000{,}000{,}000{,}000{,}000{,}000$. All of these are correct.

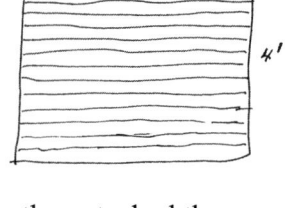

589. You put out a stack of towels for your guests to use after they have been swimming. Your kid sister took all the towels and played with them. She then stacked them but did a sloppy job. The pile was just as high but the towels weren't nicely aligned. The two stacks have the same volume because of ___Cavalieri's___ principle. (See Chapter 24, where I called it the Poker Chip principle.)

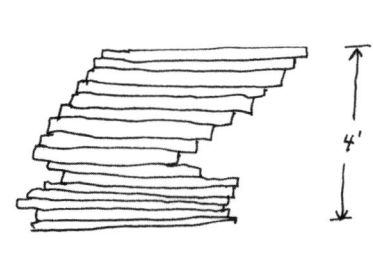

Complete Solutions and Answers | 592–595

592. Which is larger: 10^2 or 2^{10}?

10^2 is equal to $10 \times 10 = 100$.
2^{10} is equal to $2 \times 2 \times 2 \times 2 \times 2 \times 2 \times 2 \times 2 \times 2 \times 2 = 1{,}024$ So $2^{10} > 10^2$

594. It took one ounce of force to push the three-ounce box toward you at the constant speed of 1 inch per second. Do you have enough information to compute either μ_s or μ_k?

We don't have enough information to compute the constant for static friction, but $F = \mu_k N$ becomes $\qquad 1 = \mu_k(3)$.

Divide both sides by 3 $\qquad\qquad\qquad\qquad \dfrac{1}{3} = \mu_k$

What we do know about μ_s is that it must be greater than μ_k. It always takes more force to get something started than to continue to push it along at a constant speed. So $\mu_s > \dfrac{1}{3}$

595. In 1 hour Darlene can read 2 wedding novels.
In 2 hours she can read 4 wedding novels.
In 3 hours she can read 6 wedding novels.
In 4 hours she can read 7 wedding novels.
  Plot (1, 2), (2, 4), (3, 6), and (4, 7). Are these points on a straight line?

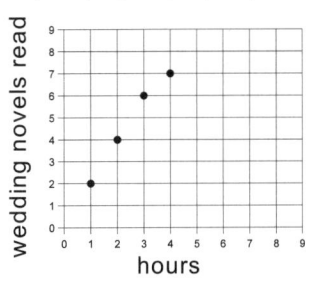

The points are not in a straight line. If they were, then Darlene would be reading at a constant rate. But this is real life. After several hours of reading, Darlene starts to get tired and reads slower.*

---

* "Reads slower" doesn't waste any words. There is no need to write "reads at a slower rate."
   In police reports the cop will often report that the suspect was driving at "a high rate of speed." Why not just say "at high speed"?

153

> 596–598 **Complete Solutions and Answers**

596. How many degrees is ∠2?

In △ AED we know that ∠1 is 30° (from the previous problem).
We know that ∠D is 90°.
We know that all three angles must add to 180°.
So ∠2 must be 60°.

597. You are now standing on the right cliff and you decide to swing back to the left cliff. Your friend Jane wants to go with you. She hangs on to you, and you both make the trip together. Will it take longer or shorter than 12 seconds to get to the left cliff?

From the second question in the *Your Turn to Play* we learned that the period is independent of the weight of the pendulum. It will take 12 seconds for you and Jane to get to the left cliff.

598. You arrived at the library. At the library snack bar you buy 5 egg sandwiches and a 3-pound onion. The whole thing weighs 10 pounds. How much does each egg sandwich weigh?

Let x = the weight of an egg sandwich.
Then 5x = the weight of the 5 egg sandwiches.
Then 5x + 3 = the weight of the whole thing.
The whole thing weighs 10 pounds.

By this time you can see that the equation is 5x + 3 = 10. That's the hard part of doing these word problems. The algebra is easy.

Subtract 3 from both sides	5x = 7
Divide both sides by 5	$x = \frac{7}{5}$
Do the arithmetic	$x = 1\frac{2}{5}$ lbs. or x = 1.4 lbs.

$\frac{7}{5} = 1\frac{2}{5}$ or   5) 7.0 = 1.4

You can submit your answer either in fractions or in decimals. Either way is fine.

154

Complete Solutions and Answers | 600–601

600. Right now, suppose you learned that 14 psi (pounds per square inch) of pressure was being applied to every square inch of your skin. Would you want to (A) yell Stop! or (B) would you like that pressure to continue?

Here is the news. Right now 14 pounds per square inch is being applied to your body. If a can of soup weighs a couple of pounds, that is like having 7 cans of soup pressing on each square inch of you.

You are probably not in pain because of this pressure. In fact, you kind of like it. To tell the truth, you would be a mess (translation: dead) if that pressure would be removed.

Here is the truth that your parents might not have told you. Air is not the same thing as nothing. Just because you can't see it or taste it or smell it, doesn't mean that it doesn't exist. Just blow against the palm of your hand and you can feel it. Air isn't very heavy. A gallon of milk weighs a lot more than a gallon of air. On the other hand, there is a lot of air stacked up for miles above you. It's been there ever since you were born. The pressure of all that air is about 14 psi against your skin.

Hey! I, your reader, don't believe that. If I had all that pressure against me, I could feel crushed. It would feel like a hippo sitting on my lap.

You are right; it would, except that your insides are pushing outward at 14 psi. The pressure of air exactly balances your outward pressure. Remove that air pressure and you would start to blow up like a balloon.

Most parents don't tell their kids about this fact of life until they are older.

When they do finally sit down to tell you about the facts of life, one of the first things they might mention is, "I know that you are under a lot of pressure."

601. Your chest of drawers now weighs 13.5 pounds. How much force will now be needed to slide it at 6 mph?

The speed doesn't matter. The weight, N, is 13.5. You know that the coefficient of friction, μ, is 0.6. (You learned that two problems ago.)

$F = \mu N$ becomes $F = (0.6)(13.5)$

$$\begin{array}{r} 13.5 \\ \times\ 0.6 \\ \hline 810 \quad 8.10 \end{array}$$

It takes 8.1 pounds to push your lightened chest across the carpet.

| 605–615 | Complete Solutions and Answers |

605. Solve $\qquad$ 3x + 7 + 9 = 19

    Do the arithmetic $\qquad$ 3x + 16 = 19

    Subtract 16 from both sides $\qquad$ 3x = 3

    Divide both sides by 3 $\qquad$ x = 1

610. For dessert after the pizza, nothing beats having some big red strawberries. Big strawberries might be 3 cm (centimeters) across. The strawberries you have are 230% larger than that. How big are they?

    There are two ways than you learned how to do "230% more than" problems.

The Hard Way: First, find out how much larger.
$$260\% \text{ of } 3 = 2.60 \times 3 = 7.80 \text{ cm}$$
Second, add that to the 3 cm you started with.
$$7.8 + 3 = 10.8 \text{ cm}$$
The Easy Way: A 260% gain means that the original size (100%) plus an extra 260%. 100% + 260% = 360%
$$360\% \text{ of } 3 = 3.6 \times 3 = 10.8 \text{ cm}$$

Since everyone knows that 5 cm is almost exactly 2 inches, these strawberries are larger than four inches—as big as your kid sister's fist.

614. The pillow began to be soaked with water. Its density increased. Its volume remained the same. Its <u>weight</u> increased.

    The definition of density (from Chapter 24) is $d = \frac{w}{v}$

which is the same as $dv = w$ (after you multiply both sides by v).

If d increases and v stays the same, then w must increase.

615. You are given 6x = 15. Find the value of x.

You start with $\qquad$ 6x = 15

Divide both sides by 6 $\qquad$ $\frac{6x}{6} = \frac{15}{6}$

Simplify both sides $\qquad$ $x = 2\frac{1}{2}$

$$6{\overline{\smash{\big)}\,15}}^{\,2\frac{3}{6}}$$
$$\underline{-12}$$
$$3$$

Complete Solutions and Answers | 619–620

619. Your kid sister said that she wanted some crayons to do a little coloring. Your mom ordered 6 cases of crayons. It came in a box that weighed 0.2 pounds. The whole thing weighed 47 pounds. How much did one of the cases weigh?

Let x = the weight of one of the cases of crayons.
Then 6x = the weight of the six cases.
0.2 = the weight of the packaging.
47 lbs. is the weight of the whole thing.

So $6x + 0.2 = 47$

Subtract 0.2 from both sides $6x = 46.8$
Divide both sides by 6 $x = 7.8$

Each case of crayons weighs 7.8 pounds.

You could probably color for six years with 7.8 pounds of crayons. I still have a box of crayons from my childhood.

620. Also on the floor is your collection of eleven colored pencils and the 9-gram pencil jar. The pencils and the jar weigh 75 grams. How much does each pencil weigh?

Your always start by letting x equal the thing you are trying to find.

Let x = the weight of a pencil
Then 11x = the weight of all eleven pencils
9 = the weight of the jar
75 = the weight of the pencils and the jar

At this point writing the equation is easy. $11x + 9 = 75$
Subtract 9 from both sides $11x = 66$
Divide both sides by 11 $x = 6$

Each pencil weighs 6 grams.

| 625–626 | Complete Solutions and Answers |

625. Suppose she drops her ball from 6 feet. Because of the loss to sound and heat, it bounces back to 3 feet. $6 \times \frac{1}{2} = 4$

Then it bounces back to 2 feet. $4 \times \frac{1}{2} = 2$

Do the arithmetic. When will it be bouncing less than $\frac{1}{100}$ of a foot?

$2 \times \frac{1}{2} = 1$

$1 \times \frac{1}{2} = \frac{1}{2}$

$\frac{1}{2} \times \frac{1}{2} = \frac{1}{4}$

$\frac{1}{4} \times \frac{1}{2} = \frac{1}{8}$

$\frac{1}{8} \times \frac{1}{2} = \frac{1}{16}$

$\frac{1}{16} \times \frac{1}{2} = \frac{1}{32}$

$\frac{1}{32} \times \frac{1}{2} = \frac{1}{64}$

$\frac{1}{64} \times \frac{1}{2} = \frac{1}{128}$ and this is less than $\frac{1}{100}$

In real life the ball might not bounce back to one-half of its previous height. It might only bounce back to 63.087% of it previous height. I chose ½ because the arithmetic would be easier for you.

Can you imagine doing . . .

$2 \times 0.63087 = 1.26174$

$1.26174 \times 0.63087 = 0.7959939138$

$0.7959939138 \times 0.63087 = 0.502168680399006$

$0.502168680399006 \times 0.63087 = 0.31680315540332091522$

Someone would have to be nuts to do that computation.

Conclusion: Your author is _____.
                           fill in one word

626. To get practical for a moment, suppose your weight gain was 0.084 pounds. What is the volume of your body?

In the previous problem we found that your weight gain would be equal to $0.08v - 0.01v$, which is $0.07v$.

$$0.084 = 0.07v$$

Divide both sides by 0.07 $\frac{0.084}{0.07} = v$ Doing the arithmetic $v = 1.2$ ft^3

158

Complete Solutions and Answers — 639–641

639. On Mars it is true that 1000 meters = 1 kilometer.
On earth 1 kilogram ≈ 2.205 pounds. Is that true on Mars?

    Kilograms are a measure of mass. It is the amount of "stuff" in some object. If you kid sister has a mass of 15 kilograms on earth, she will have a mass of 15 kilograms on Mars.

    Pounds are a measure of force. You put a 1 kilogram weight on a scale here on earth and it will weigh 2.205 pounds. You put that same 1 kilogram mass on a scale on Mars (which is a smaller planet than earth) and it will weigh less than 2.205 pounds.

    If you put an elephant on a very small rock with very little gravity—small planets are not called planets—it might only weigh one ounce.

    C. C. Coalback once ran a weight-reduction clinic. He guaranteed that anyone who paid their $6,250 entrance fee would lose half their weight in 2 weeks. He shipped his customers to the moon where they lost over half their weight, but didn't lose any of their mass.

640. If it takes 25 pounds of force to keep the duck moving at 5 mph, how many pounds of force will it take to move him at a constant 15 mph?

    Friction and force are independent of speed. It will take 25 pounds of force to keep him going at any constant speed. (Of course, it takes more effort to get him up to that faster speed, but once he is there, 25 pounds of force will keep him going at any constant speed.)

641. You are given $8x = 7$. Find the value of x. Express your answer both as a fraction and as a decimal.

You start with	$8x = 7$
Divide both sides by 8	$x = \dfrac{7}{8}$

To convert $\dfrac{7}{8}$ into a decimal, you divide.

$\dfrac{7}{8} = 0.875$ which as a percent is 87.5%

A free bonus!

```
       0.875
    8)7.000
     - 64
       60
     - 56
       40
     - 40
```

159

642–644 Complete Solutions and Answers

642. You lift 13 ounces of fish 24 inches upward. How many foot-pounds of work did you use in lifting them?

We need to convert everything to pounds and feet.

$$\frac{13 \text{ oz}}{1} \times \frac{1 \text{ lb.}}{16 \text{ oz}} = \frac{13}{16} \text{ pounds}$$

$$\frac{24 \text{ in}}{1} \times \frac{1 \text{ ft}}{12 \text{ in}} = 2 \text{ feet}$$

Work = Force × distance
W = Fd becomes

$$W = \frac{13}{16} \times \frac{2}{1} = \frac{13}{8} = 1\frac{5}{8} \text{ ft-lbs.}$$

643. Your kid sister's sticker collection was world-famous. Before her alligator ate all her stickers, she owned 8,398,200,335 stickers. Now she owns 0 stickers. Is the number of stickers owned a discrete or a continuous variable?

You can own 8 stickers or you can own 9 stickers, but you can't own every number between 8 and 9. You can't own 8.986906379 stickers. It is a discrete variable.

644. The density is 0.8 pounds per cubic inch. If you dove 10 feet under the surface of the lake, what would be the pressure?

Pressure of a liquid (or gas) is equal to its density times the height.

p = dh becomes

$$p = \frac{0.8 \text{ lbs.}}{\text{in}^3} \times \frac{10 \text{ feet}}{1}$$

10' = 120"

$$p = \frac{0.8 \text{ lbs.}}{\text{in}^3} \times \frac{120 \text{ in}}{1}$$

Doing the arithmetic

$$p = 96 \text{ lbs./in}^2$$

Discussion: Since the atmospheric pressure is about 14.7, the pressure of the Sluice would be $\frac{96}{14.7}$ or about six and a half times as much. In physics that is called 6.5 atmospheres.

Complete Solutions and Answers | 655–657

655. In the hallway the hose weighed 4.8 pounds and μ_k was 0.7. How much did she have to pull on the hose to keep it moving?

$F = \mu_k N$ becomes $\qquad\qquad F = (0.7)(4.8)$

Doing the arithmetic $\qquad\qquad\qquad\quad F = 3.36$ pounds

```
    4.8
  × 0.7
   336     3.36
```

656. You give your kid sister a 1.7-pound slice of your 6-pound pizza. How much is left for you?

```
    6.                      6.0
  − 1.7      becomes      − 1.7
                            4.3
```

You have 4.3 pounds. You love your sister and would do most anything for her. At the same time, she sometimes drives you nuts, and you hate what she does. You definitely don't love her alligator.

657. When he was standing on his board, μ was 0.6. He is now lying in the sand holding his board. Will μ change?

The weight of the surfer and his board (140 pounds) has not changed. You might have remembered back to Chapter 1 that when Fred pulled his toy safe across his desktop it didn't matter whether it was upright or on its side. The amount of pull was the same. And μ was the same.

But the case with the surfer is different. μ measures the friction between two surfaces that are touching. In the case of the toy safe, it was the metal of the toy safe touching the desktop—both when the safe was upright and when it was on its side.

But when the surfer was upright, it was his slick board touching the sand. When he fell over, it was his hairy chest touching the sand. Everyone knows that hairy chests are harder than slick surfboards to drag over sand. μ depends on which two surfaces are touching. μ probably increased.

| 658–661 | Complete Solutions and Answers |

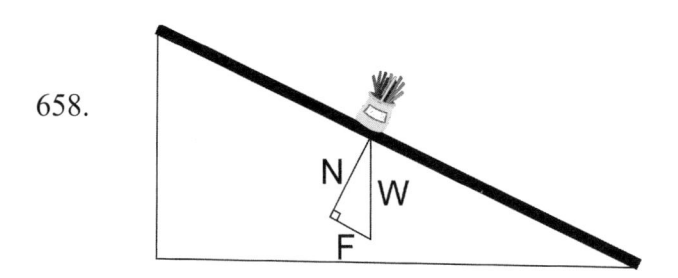

658.

659. From this bunch of measurements I conclude that the sum of the angles in any triangle is equal to 180°. Is this an example of inductive or deductive reasoning?

    Whenever we go from experiments, observations, or trials to a conclusion, we are using inductive reasoning.

660. How many degrees is ∠3?

    Since the vector **N** is at right angles to segment BEA, we know that ∠2 + ∠3 must be 90°

    We know that ∠2 is 60° (from the previous problem).

    So ∠3 must be 30°.   60° + 30° = 90°

661. Ohm's law is V = IR, where V is the voltage, I is the amperage, and R is the resistance. Which of these are true?
1) V = RI   2) V/R = I   3) R = V/I   4) IR = V

1) V = RI is true. That is because IR is always equal to RI. When you multiply two numbers together, it doesn't matter which order you multiply them. 7 × 8 is equal to 56. So is 8 × 7.

2) V/R = I is true. If we start with Ohm's law, V = IR and divide both sides by R, we will get V/R = I

3) R = V/I is true. First, start with Ohm's law   V = IR
        Divide both sides by I   V/I = R
        If a = b then b = a   R = V/I

4) IR = V is true. In algebra, we will call the fact that if a = b, then b = a, the symmetric law of equality. You know that is true. Now you have a name for it.

Complete Solutions and Answers | 673–675

673. Why didn't the spring stretch 20 feet?

   The ceilings in most American bedrooms are 8 feet tall. Some are 10 feet. When the alligator hit the floor, the spring isn't going to stretch any farther.

674. What is the volume of your stack of pizzas?

   The area of each pizza is 75 square inches.
   The height of the stack is 8 feet.
   We either need to work in feet or in inches. If I work in feet, then I'll have to convert 75 square inches into feet and the work will look like what I've stuck in the box.

$$\frac{75 \text{ sq in}}{1} \times \frac{1 \text{ sq ft}}{144 \text{ sq in}} = \frac{75}{144} \text{ sq ft}$$

   I don't want to work in fractions or in decimals if I can avoid it. In this case, decimals would be even worse than fractions. Everyone knows that $\frac{75}{144} \approx 0.5208333333333333333333333333333333$.

   So I'll convert everything into inches. $\frac{8 \text{ ft}}{1} \times \frac{12 \text{ in}}{1 \text{ ft}} = 96 \text{ in}$

Cavalieri's principle says that the volume of a stack is the area of any of the poker chips times the height of the stack.

   $V_{\text{stack of pizzas}} = 75 \text{ in}^2 \times 96 \text{ inches} = 7{,}200 \text{ in}^3$.

675. If you were dragging it at 4 mph and then later you were dragging it at 8 mph, it took an extra bit of force to get it moving at the faster speed. That extra bit of force is called inertia.

   The *tia* at the end of *inertia* is pronounced sha.
   That is like the *tion* at the end of *nation* is pronounced shun.
   But the *ture* at the end of *nature* is pronounced chur.
   None of these are like the *t* in *cat*, *rat*, or *bat*.
   And the *t* in *soften* isn't pronounced at all.

English teachers have a lot more work to do than math teachers.

676–678 | Complete Solutions and Answers

676. Does the buoyancy of the pizza in the bath water depend on how dense your pizza is?

To repeat part of the answer from the previous problem: The buoyancy of an object is given by the formula buoyancy = dv where d is the density of the water (or other fluid) and v is the volume of the object that is submerged.

d is the density of the water, *not of the pizza.*

How much the water is pushing the pizza upward (not *upwards*) depends on how much volume the pizza displaces, not how dense it is.

Of course, the density of your pizza will determine whether it sinks or floats, but that is the topic of the next chapter (*Chapter 26: Why Things Sink*).

677. You have gone duck-nuts. The only thing you buy are more ducks. Is the number of ducks you buy proportional to the amount of money you have?

Yes. If you have, say, $10, you might buy x ducks. If you have $20, you would buy 2x ducks.

If you had a million dollars

678. The first thing she did was to take your scale. You had left it in her room so she figured it was hers. She lifted that 4-pound scale 16 inches in the air and carried it into her bathroom. How much work (in foot-pounds) did she do?

Work is equal to force times distance.
W = Fd becomes   4 lbs. × 16 inches
Since we want an answer in ft-lbs.   4 lbs. × 1⅓ feet
Doing the arithmetic   5⅓ ft-lbs

Here is the arithmetic: $\frac{16 \text{ in}}{1} \times \frac{1 \text{ ft}}{12 \text{ in}} = \frac{16}{12} = 1\frac{4}{12} = 1\frac{1}{3}$ feet

and $\frac{4}{1} \times 1\frac{1}{3} = \frac{4}{1} \times \frac{4}{3} = \frac{16}{3} = 5\frac{1}{3}$

164

Complete Solutions and Answers | 679–681

679. In this circuit a 9-volt battery will create a current of 10 amperes. (The current will depend on the resistance.) If I replace the battery with a 110-volt power source, what will the current be? Round your answer to the nearest amp.

The conversion factor will be either $\dfrac{9 \text{ volts}}{10 \text{ amps}}$ or $\dfrac{10 \text{ amps}}{9 \text{ volts}}$

$$\dfrac{110 \text{ volts}}{1} \times \dfrac{10 \text{ amps}}{9 \text{ volts}}$$

$$= \dfrac{110 \text{ v\cancel{olts}}}{1} \times \dfrac{10 \text{ amps}}{9 \text{ v\cancel{olts}}}$$

$$= \dfrac{1100}{9} \text{ amps}$$

$$\quad 9\overline{)1100.0}\ \ 122.2$$

$$\doteq 122 \text{ amps}$$

680. Your gardener will plant your 3,750-square-foot elliptical lawn. It will cost $40 for each 250 square feet he plants. What will be the total cost? (Use a conversion factor.)

The conversion factor will be either $\dfrac{250 \text{ ft}^2}{\$40}$ or $\dfrac{\$40}{250 \text{ ft}^2}$

$$\dfrac{3750 \text{ ft}^2}{1} \times \dfrac{\$40}{250 \text{ ft}^2} = \dfrac{3750 \cancel{\text{ft}^2}}{1} \times \dfrac{\$40}{250 \cancel{\text{ft}^2}} = \$600$$

The arithmetic: $\begin{array}{r} 3750 \\ \times\ \ 40 \\ \hline 15000 \end{array}$ $250\overline{)15000.}\ \ 600.$

681. In the west wing of your dream house are 12 bedrooms. They are all the same size. A painter says that he could paint all the rooms for $5,000. Of that $5,000, $3,524 would be the cost of the paint and the rest is his labor cost. How much would be the labor cost for each of the bedrooms?

You always start by letting x equal the thing you are trying to find.
Let x = the labor cost for one bedroom
Then 12x = the labor cost for 12 bedrooms
3524 = cost of the paint

Now the equation is easy to write. $\quad 12x + 3524 = 5000$

Subtract 3524 from both sides $\quad 12x = 1476$

Divide both sides by 12 $\quad x = \$123$

709–711 Complete Solutions and Answers

709. An ellipse—2 feet by 1 foot. To the nearest tenth of a square foot, what is its area?

If the major axis is 2 feet and the minor axis is 1 foot, then a = 1 and b = ½.

$Area_{ellipse} = \pi ab = \pi(1)(½) = \pi/2$. We want $\pi/2$ to the nearest tenth of the square foot. $\pi/2 \approx \dfrac{3.1416}{2} = 1.5708 \doteq 1.6$ square feet

≈ means approximately equal to
≐ means equals after rounding

A 1.6 square foot pizza makes a nice snack.

710. Your kid sister holds out here little chubby hand and your mom plops the 18-pound pizza on it. Her hand has the same area as a 2"×3" rectangle. What is the pressure on her hand?

Pressure is defined as $\dfrac{Force}{Area}$ or $P = \dfrac{F}{A}$

We know that the force is 18 pounds, which is the weight of the pizza.

The area of her hand is equal to $A_{rectangle} = \ell w = 2 \times 3 = 6$ in².

$P = \dfrac{F}{A}$ becomes $\dfrac{18 \text{ lbs.}}{6 \text{ in}^2} = 3$ lbs./in² (or 3 psi)

711. The alligator weighs 18 pounds. Can you find μ for the alligator and the carpet in your bedroom?

Since we can't look up in a book the value of μ for her alligator and the carpet in your bedroom, the only way we can find μ is using the formula $F = \mu N$.

To find μ, we need to know both F and N. All we know is that the weight of the alligator, N, is 18 pounds.

If we had been told how much force, F, your kid sister needed to drag her 'gator across your bedroom carpet, then we could find μ.

But with what we know, we can't find μ.

Complete Solutions and Answers | 712–714

712. $E = mc^2$ means that a little bit of matter is equal to a huge amount of energy. Suppose m were equal to, say, 5. What would E be equal to?

For this problem, to make the arithmetic a little easier, suppose that c is equal to 300,000,000 rather than 299,792,458.

$E = mc^2$ becomes $E = 5 \times 300{,}000{,}000 \times 300{,}000{,}000$

Here's the arithmetic:
```
          300,000,000
       ×  300,000,000
     90,000,000,000,000,000
```

```
     90,000,000,000,000,000
   ×                     5
    450,000,000,000,000,000
```

Turning 5 units of matter into energy would yield four hundred and five quadrillion units of energy.

Of course, you can take four hundred and fifty quadrillion units of energy and turn it into 5 units of matter.

million	1,000,000
billion	10^9
trillion	10^{12}
quadrillion	10^{15}

If you could speak loudly enough, you could create grains of sand, trees, oceans, worlds, stars, the whole universe. That's what physicists say.

713. What did each of these three letters of p = dh stand for?

p is the pressure
d is the density of the liquird
h is the height (or depth) of the object below the surface

714. How much more would 5 ounces of silver weigh than 5 ounces of chocolate? (Use one avoirdupois ounce ≐ 28.3 grams and one troy ounces ≐ 31.1 grams.)

5 ounces of chocolate = 5×28.3 = 141.5g   (Don't use your calculator!)
5 ounces of silver = 5×31.1 = 155.5g

The difference is (155.5 – 141.5) 14 grams.

| 715–717 | **Complete Solutions and Answers** |

715. You never see a duck go ice skating on a carpet. It's just too much work. That illustrates the fact that friction is dependent on . . .

    A) the area of contact between the surfaces

    B) the speed at which he's skating

    C) which surfaces are involved (in this case, between the bottom of his skates and the carpet)

    A) We know that the area of contact doesn't change the friction.

    B) We know that skating at any constant speed doesn't affect the friction.

    C) But there is a lot more friction skating on carpet than on ice. It does matter which surfaces are involved. (Besides being much more work to use your ice skates on the carpet in your living room, there is another reason. If you are not sure what that second reason is, ask your mother.)

716. Plot the points (1, 3) and (20, 4).

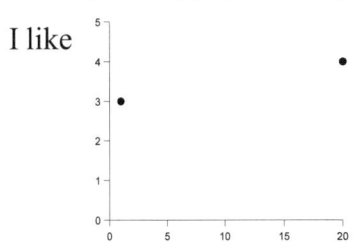

I like

You might like

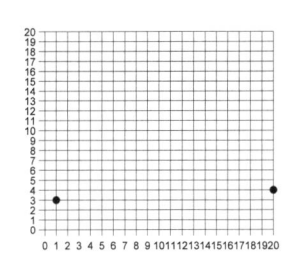

717. Is the number of bulldozers I own a discrete or a continuous variable?

    I could own 3 bulldozers. I could own zero bulldozers. But I couldn't own 1.86396 bulldozers. It is a discrete variable.

This isn't possible.

Complete Solutions and Answers | 718–719

718. Convert 1.6 square feet into square inches. Use a conversion factor. 144 square inches = 1 square foot.

The conversion factor will be either $\frac{144 \text{ sq in}}{1 \text{ sq ft}}$ or $\frac{1 \text{ sq ft}}{144 \text{ sq in}}$.

$$\frac{1.6 \text{ sq ft}}{1} \times \frac{144 \text{ sq in}}{1 \text{ sq ft}} = 230.4 \text{ square inches}$$

I chose the $\frac{144 \text{ sq in}}{1 \text{ sq ft}}$ so that the square feet would cancel. I wanted to get rid of square feet and arrive at square inches.

719. A 6-pound scuba tank stretches the rubber band 8 inches. How far would it stretch if you add a 4-pound scale?

Hooke's law says that the stretch, x, is proportional to the amount of force, F, applied to the spring. $F = kx$

This is a two-step problem. First, we need to find k. Then we can find the stretch, x.

$F = kx$ becomes	$6 = k8$
Putting the number in front of the letter	$6 = 8k$
Divide both sides by 8	$\frac{6}{8} = k$
Doing the arithmetic	$\frac{3}{4} = k$

Now to find the stretch after the 4-pound scale is added to the 6-pound scuba tank.

$F = kx$ becomes	$4 + 6 = \frac{3}{4}x$
Doing the arithmetic	$10 = \frac{3}{4}x$
Dividing both sides by $\frac{3}{4}$	$13\frac{1}{3} = x$

The spring stretches 13⅓ inches.

Here is the arithmetic: $10 \div \frac{3}{4} = \frac{10}{1} \times \frac{4}{3} = \frac{40}{3} = 13\frac{1}{3}$

$$3)\overline{40} \quad \begin{array}{r} 13 \text{ R1} \\ \underline{3} \\ 10 \\ \underline{9} \\ 1 \end{array} = 13\frac{1}{3}$$

| 720–722 | Complete Solutions and Answers

720. How many degrees is ∠4?

We know that the angles in the little triangle must add to 180°. Namely, ∠3 + ∠4 + 90° = 180°.

We know that ∠3 is 30° (from the previous problem).

So ∠4 must be 60°.

Discussion: We know that the angles in the big triangle (△ABC) are 30°, 60°, and 90°. We have found that the angles in the small triangle are also 30°, 60°, and 90°.

If we had started with ∠B equaling 70°, we would have found that the angles in both the big and small triangles would be 70°, 20°, and 90°.

In any event, the angles in the big triangle will equal angles in the small triangle. So the two triangles are always similar.

Similar triangles have sides that are in the same ratio.

Translation: In the big triangle, $\frac{rise}{run}$ will equal $\frac{F}{N}$ in the small triangle.

Kingie was right. We can find the coefficient of static friction, μ_s, without having to weigh the object, N, or find out how much force, F, is needed to start it moving.

721. She popped your kid sister on top of you and dragged the pair of you out into the hallway where there was fresh air. If μ_k was 0.54, what would it have been if she had just dragged you?

μ_k is the coefficient of friction between two surfaces. It does not change if the force between the surfaces (the weight) changes.

722. You are in your bedroom with the door shut. It's 11 p.m. You hear the familiar bounce, bounce, bounce, bounce, bounce, bounce, bounce, bounce, bounce, bounce, bounce, bounce, bounce, bounce, bounce. For the last week your kid sister has been out in the hallway bouncing her big rubber ball in the middle of the night.

How certain are you that she's out there in the hallway bouncing her ball?

You are probably 99.9% certain, but not 100% certain. You are like a scientist who has been doing a bunch of experiments. Every time you have heard bounce, bounce, bounce, bounce, bounce, bounce, bounce, bounce,

Complete Solutions and Answers | 723

bounce, bounce, bounce, bounce, bounce you have gone out into the hallway and seen her bouncing her stupid ball.

The first time you heard the bounce, bounce, bounce you had a hunch that it was her. By the fourth time, your hunch had turned into a law. You "knew" that it was her.

Imagine your surprise when you found that she was sound asleep in her bed and that her alligator was out in the hallway pounding his tail against the floor. It made a sound just like a rubber ball bouncing.

bounce, bounce

723. The pizza weighs 4½ pounds. You take your fork and push it an inch to the left. How much force will you need to apply?

First, we find μ_s between the pizza and your Pizza Ramp.

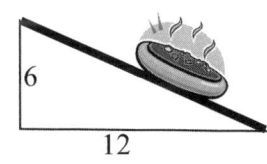

The rise is 6 and the run is 12. According to Kingie, μ_s = rise/run = 6/12 = $\frac{1}{2}$

Second, we have a 4½ pound pizza and μ_s is ½.
$F = \mu_s N$ becomes $\qquad F = (½)(4½)$
Doing the arithmetic $\qquad\qquad F = 2¼$

$$\frac{1}{2} \times 4\frac{1}{2}$$
$$= \frac{1}{2} \times \frac{9}{2}$$

$$\begin{array}{r} 2 \text{ R } 1 \\ 4)\overline{9} \\ -8 \\ \hline 1 \end{array}$$

$$= \frac{9}{4}$$

$$= 2¼$$

It takes two and a quarter pounds to push that pizza.

171

| 724–726 | **Complete Solutions and Answers** |

724. After you have loaded all the things that were in your chest of drawers onto that shelf, the shelf starts to tip. The first thing to start to slide is your Christina Rossetti book. Find the coefficient of static friction between the book and the shelf.

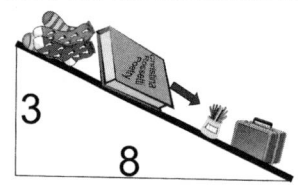

$$\mu_{\text{static friction}} = \frac{\text{rise}}{\text{run}} = \frac{3}{8}$$

or, if you like decimals, $\mu_{\text{static friction}} = 0.375$

725. Graph the points (2, 1) and (5, 3). What is the slope of the line that connects those two points?

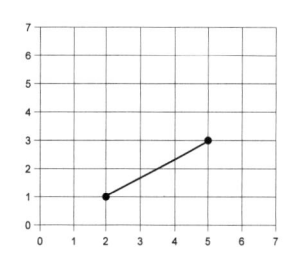

becomes ⇨

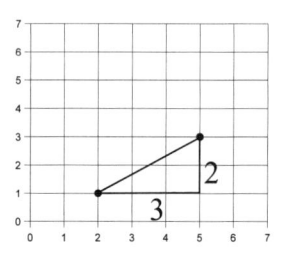

$$\text{Slope} = \frac{\text{rise}}{\text{run}} = \frac{2}{3}$$

726. You grab a broom and push the 0.08 pounds of the mess 12 feet across the room and into a corner so that no one will step on it. The coefficient of friction between the mess and the floor is 0.5. How much work did you do?

Work equals force times distance   W = Fd
force equals the coefficient of friction times the weight   F = μN

F = μN becomes   F = (0.5)(0.08)
Doing the arithmetic   F = 0.04
W = Fd becomes   W = (0.04)(12)
Doing the arithmetic   W = 0.48 ft-lbs

Complete Solutions and Answers | 740–745

740. The density of the oil is 50 lbs/ft^3. The pool measures 20' × 30' × 8'. What is the weight of the oil?

The volume of a box is length times width times depth.

$V_{box} = \ell wd$ becomes $V_{swimming\,pool} = (30)(20)(8)$
$V_{swimming\,pool} = 4{,}800$ ft^3

The definition of density (from Chapter 24) is density = $\dfrac{weight}{volume}$

$d = \dfrac{W}{V}$ becomes $50 = \dfrac{W}{4{,}800}$

Multiply both sides by 4800 $50(4800) = w$
Doing the arithmetic $240{,}000 = w$

The weight of the oil in your pool is 240,000 pounds

744. Your kid sister removed the chicken from the soup. All that was left was the bowl and some warm water. Did this change μ?

No. Mu depends only on the two surfaces and not on how hard the surfaces are pressing against each other.

745. Since we know that the jar is not moving, we know that the coefficient of static friction, μ_s, between the jar and the shelf is greater than what number? (This takes us back to Chapters 13 in which Kingie found a formula that flabbergasted Fred.)

That formula was $\mu_s = \dfrac{rise}{run}$

The rise/run in this diagram is $\dfrac{3}{8}$

If the jar were just starting to move with this slope of $\dfrac{3}{8}$ then we could say that μ_s is equal to $\dfrac{3}{8}$

Since it is not moving, μ_s must be greater than $\dfrac{3}{8}$

| 750–751 | Complete Solutions and Answers |

750. Glob, 6 pounds. Alligator, 18 pounds. Sister, 24 pounds. How much force will be needed to pull the glob + alligator + sister at a constant speed toward the garbage can? From the previous problem we know that it takes 10 pounds to move glob + alligator at a constant speed.

The solution to this problem takes a couple of steps. We will use $F = \mu N$ twice. As a general rule, a two-step problem is about four times as hard as a one-step problem.

From it takes 10 pounds to move glob + alligator at a constant speed:

$F = \mu N$ becomes	$10 = \mu(6 + 18)$
Doing the arithmetic	$10 = \mu 24$
Putting the number in front of the letter	$10 = 24\mu$
Dividing both sides by 24	$\frac{5}{12} = \mu$

In this case, I like $\frac{5}{12}$ much better than working in decimals. $\frac{5}{12}$ seems better than $0.416666666\ldots$ or $0.41\overline{6}$

Now, knowing the value of μ, and that N = glob + alligator + sister:

$F = \mu N$ becomes	$F = \frac{5}{12}(6 + 18 + 24)$
Doing the arithmetic	$F = \frac{5}{12} \times 48$
Doing more arithmetic	$F = 20$

It will take 20 pounds of force to slide all three toward the garbage can.

751. $2,000 is what percent of $8,000?

When you don't know both sides of the *of* you divide the number closest to the *of* into the other number.

```
            0.25
      8000) 2000.00
            16000
             40000
             40000
```

$0.25 = 25.\% = 25\%$

The sales from one-quarter of one of Kingie's paintings would pay for the repair.

174

Complete Solutions and Answers | 754–756

754. You put 5 of these big strawberries on a 0.7 kg (kilogram) plate. The whole thing (strawberries + plate) weighs 3.7 kg. How much does each strawberry weigh?

Let x = the weight of one strawberry.
Then 5x = the weight of the 5 strawberries.
Then 5x + 0.7 = the weight of the strawberries plus the plate.
We know that the whole thing weighs 3.7 kg.

$$\text{The equation becomes} \quad 5x + 0.7 = 3.7$$

Subtract 0.7 from both sides $\quad 5x = 3$

Divide both sides by 5 $\quad x = 0.6$

$$5 \overline{) 3.0} \quad 0.6$$

Each of those giant strawberries weighs 0.6 kilograms.

755. We know that the period of a pendulum is proportional to the square root of the length. The period is proportional to $\sqrt{\text{length}}$.

That means that if I make the rope four times as long, the period will be twice as long. $\sqrt{4} = 2$.

How long should the rope be in order to make the period three times as long?

If the rope is nine times as long, the period will triple. $\sqrt{9} = 3$.

756. You can eat 20% of an egg sandwich in a minute. (Translation: In one minute you can eat 0.2 egg sandwiches.)

How long will it take you to eat your 5 egg sandwiches? Use a conversion factor.

We know that 1 minute = 0.2 egg sandwiches so the conversion factor will be either $\frac{1 \text{ minute}}{0.2 \text{ egg}}$ or it will be $\frac{0.2 \text{ egg}}{1 \text{ minute}}$

$$\frac{5 \text{ egg}}{1} \times \frac{1 \text{ minute}}{0.2 \text{ egg}} = \frac{5 \text{ e̶g̶g̶}}{1} \times \frac{1 \text{ minute}}{0.2 \text{ e̶g̶g̶}}$$

$= \frac{5}{0.2}$ minutes = 25 minutes to finish the 5 egg sandwiches.

| 757–760 | **Complete Solutions and Answers**

757. You carry that pair of fish (13 ounces) clutched to your chest as you transport them to their final resting place. They stay 4 feet off the ground. You walk 11 feet to the wastebasket. How much work would a physicist say that you had done in carrying those fish?

Physicists are weird. If you were to carry those precious pets of yours a million miles (and kept them 4 feet off the ground), they would claim that you hadn't done any work at all. W = Fd, where d is the distance lifted, means that W = 0 when d = 0.

758. The pillow will begin to sink when its density exceeds what number?

As we mentioned three problems ago in #200, in general, things will float if their density of less than the liquid they are in. They will sink if their density is greater than the liquid they are in.

So the pillow will begin to sink when its density is greater than 0.59 oz./in^3.

759. Sluice is also the world's most dense soft drink since it is 99% sugar and 1% water. She gives your kid sister a 34-foot straw. What is going to happen?

Air pressure will support a column of water that is 34 feet tall. Sluice is denser than water. For a given volume, Sluice will weigh more than water. The column of Sluice will be shorter than a column of water. Your kid sister won't get any Sluice.

760. The front door to your bedroom is three feet wide and six feet tall. She drew the biggest ellipse that would fit on your door and painted it in. What is the area of that ellipse? Use 3 for π in this problem.

The ellipse is 6 feet tall and 3 feet wide. For English majors we say that the **major axis** of the ellipse is 6 feet, and the **minor axis** is 3 feet. Cutting those in half, we get the values for a and b.

$A_{ellipse} = \pi ab$ becomes A = (3)(3)(1.5)

Doing the arithmetic   A = 13.5 square feet
Or, if you like fractions   A = 13½ square feet

Complete Solutions and Answers 761–762

761. Captain Duck was going 8 mph (miles per hour). How long would it take him to go 300 feet. (5,280 feet = 1 mile) Use conversion factors.

    I have to convert either feet into miles or convert miles into feet. There will be less fractions if I convert miles into feet, and hours into minutes.

$$\frac{8 \text{ miles}}{1 \text{ hour}} \times \frac{5280 \text{ feet}}{1 \text{ mile}} \times \frac{1 \text{ hour}}{60 \text{ minutes}} = 704 \text{ feet/minute}$$

d = rt becomes   300 feet = (704 feet/minute)t

Divide both sides by 704 $\frac{300}{704}$ minutes = t

$\frac{300}{704}$ minutes is fairly ugly looking. Let's convert it to seconds.

$$\frac{300 \text{ minutes}}{704} \times \frac{60 \text{ seconds}}{1 \text{ minute}} = \frac{18000}{704} \text{ seconds.}$$

If I do the division, that is roughly 26 seconds.

```
        25.5
704) 18000.0
     1408
     3920
     3520
     4000
     3520
```

762. Any physics handbook will tell you that 1 pound ≈ 4.448 newtons.
    Using a conversion factor, translate a push of 5 pounds into the metric system.

    The conversion factor will be either $\frac{4.448 \text{ newtons}}{1 \text{ lb.}}$ or it will be $\frac{1 \text{ lb.}}{4.448 \text{ newtons}}$

$$\frac{5 \text{ lbs.}}{1} \times \frac{4.448 \text{ newtons}}{1 \text{ lb.}}$$

```
  4.448
×     5
 22240     22.240
```

= 22.24 newtons

177

| 763–765 | Complete Solutions and Answers

763. When that alligator is not tied up or hanging from the ceiling, your sister lets it run around in her bedroom. Sometimes it is 7 feet from her bedroom door. Sometimes it is 1 foot from her bedroom door. Is the distance that the alligator is from her bedroom door a discrete or a continuous variable?

First, notice that the alligator will never be 50 feet from her bedroom door as it crawls around in your sister's bedroom. Her bedroom is not that large. *That doesn't matter when trying to figure out whether it is a discrete or continuous variable.*

When does matter is the question of whether the variable can take on every value between 7 and 1. If a variable is continuous and can have values of 7 and 1, then it must be able to have any value between 7 and 1.

Suppose the alligator is 7 feet from her door. You knock on that door. The alligator crawls to 1 foot from the door. (It is looking to see if there is someone at the door that is good to eat.) In making that crawl from 7 to 1 it will have at some instant in time been 5.36 feet from the door. There is no way that it could be 5.35 feet and then be 5.37 feet without passing through 5.36 feet from the door. It is a continuous variable.

764. Your kid sister will have a house that has 3,000 square feet in it. Your house will be 260% larger than hers. How many square feet will be in yours?

There are two ways to do "15% larger than" problems.

The hard way: 260% of 3,000 is 7,800. Your house will be 7,800 ft² larger than hers. So your house will be 3,000 + 7,800 = 10,800 ft².

The easier way: A 260% gain means the original 100% plus an extra 260%. 260% + 100% = 360%. 360% of 3,000 is 10,800 ft².

765. You don't want the oil to go to waste. You have a rusted combination lock. You toss it into the pool. It sinks to the bottom. What is the pressure on that lock?

pressure = density times height
p = dh becomes
p = (50 lbs/ft³)(8 ft) = 400 lbs/ft²

Complete Solutions and Answers | 797–799

797.

$20 = 5x$	$8y = 16$	$22w = 11$	$60 = 100v$	$4\mu = 12$
$4 = x$	$y = 2$	$w = \frac{11}{22}$	$\frac{60}{100} = v$	$\mu = 3$
		$w = \frac{1}{2}$	$\frac{3}{5} = v$	

798. 1.7 is what percent of 6? Round your answer to the nearest percent.

1.7 = ?% of 6. We don't know both sides of the *of*, so we divide the number closest to the *of* into the other number. (This is what was taught in *Life of Fred: Decimals and Percents*.)

$$\begin{array}{r} 0.283 \\ 6\overline{)1.700} \\ \underline{12} \\ 50 \\ \underline{48} \\ 20 \\ \underline{18} \\ \end{array}$$

$0.283 = 28.3\% \doteq 28\%$

You gave your kid sister 28% of your pizza. You are nice.

799. You hang your 5 ounces of chocolate on a 4-foot string attached to the ceiling and start it swinging. You measure the period (the time it takes to make a complete trip back and forth).

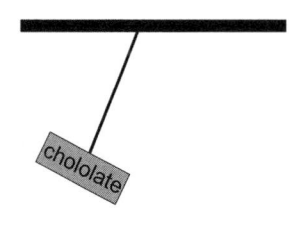

As it swings over near you, you take a bite out of the bar. Does this increase or decrease the period?

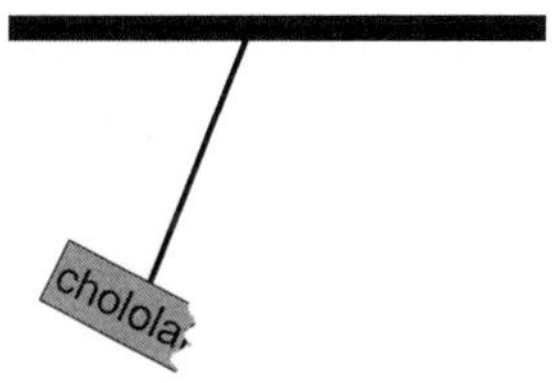

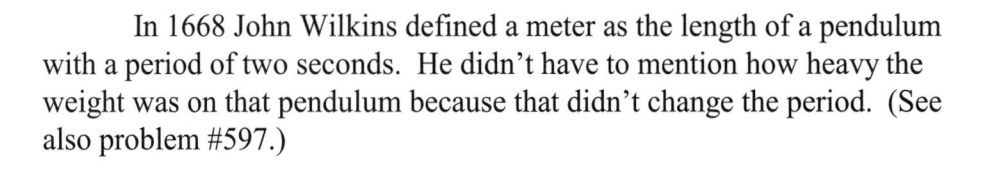

In 1668 John Wilkins defined a meter as the length of a pendulum with a period of two seconds. He didn't have to mention how heavy the weight was on that pendulum because that didn't change the period. (See also problem #597.)

| 805–811 | **Complete Solutions and Answers** |

805. Pulling one board behind him at a constant speed required 6 pounds of force.

Some kids tied a second board behind the first.

How much force did he need to apply now?

μ did not change because the surfaces did not change. It was still surfboards touching sand.

The weight doubled.
If we start with F = μW,
  if we keep μ the same,
    if we double W,*
      then F will double.
He will need to apply 12 pounds of force.

809. You are happily dragging your backpack down the street at 8 mph. Your kid sister grabs it and your speed slows to 7 mph, then 6 mph, then 5 mph, then 4 mph. The force that she was exerting on your backpack is called inertia.

Inertia is defined to be the force needed to change an object's speed or direction. It can speed things up or slow them down.

811. Your kid sister was trying to be helpful. One of the 4,999 guest was smoking. She grabbed that cigarette and threw it away. She threw it into your swimming pool—the pool filled with oil. Water doesn't burn. Oil does. A huge fire and explosion filled the air with orange flames and black smoke. All the guests were hurt, and your dream house burned to the ground. She took the combination lock, which was unharmed by the fire, and walked down your driveway.

---

* "Double w" is hard to say. Saying "w" sounds almost like saying, "double u." Did you ever wonder about that?
  If I take a couple of u's   uu
  and smash them together   ɯ
  what do you think I get?

180

Complete Solutions and Answers — 816

816. The formula for the circumference of a circle is $C = 2\pi r$.

Given an ellipse, can you give the formula for the distance around the edge (the perimeter)?

Going from the area of an ellipse, $A = \pi ab$ to the area of a circle, $A = \pi r^2$, was super easy.

In real life, you don't always know whether a problem will be easy or hard.

Your mother can't seem to open a jar of pickles. She hands it to you and says, "You're big and strong. Please open this for me."

First, you just use your muscles. No luck. That lid is on tight.

Then you tap the jar gently against the counter. That will sometimes loosen a lid. No luck.

Then you run the jar under hot water. That will sometimes work. No luck.

You tap it gently with a hammer. (not recommended!) No luck.

It's time to get serious. Dynamite will always work. Boom! It worked. The only problem is that the jar is completely shattered and every pickle has glass shards in it.

So what's the formula for the distance around the edge of an ellipse?

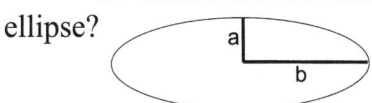

It is *approximately equal* to $\quad \text{Perimeter}_{\text{ellipse}} \approx 2\pi\sqrt{\tfrac{1}{2}(a^2 + b^2)}$

It is *exactly equal* to $\quad \text{Perimeter}_{\text{ellipse}} = 4a \int_{\theta=0}^{\pi/2} \sqrt{1 - k^2 \sin^2\theta}\, d\theta$

where $k = \dfrac{\sqrt{a^2 - b^2}}{a}$ Most readers will not have guessed this.

817–820 | Complete Solutions and Answers

817. Find x. $50 = 80x$

Divide both sides by 80 $\dfrac{50}{80} = x$

Simplify $\dfrac{5}{8} = x$

818. If your driveway is 1,200 feet long and she can walk at 3 feet per second, how long will it take for her to walk up your driveway? Give you answer in minutes and seconds.

    distance equals rate times time.
    Let t = the number of seconds it takes her to walk up your driveway

    d = rt becomes   1200 = 3t
    Divide both sides by 3   400 = t

It will take her 400 seconds.
To convert seconds into minutes, we divide by 60.

```
      6 R 40
60) 400              6 minutes and 40 seconds
    360
     40
```

819. Solve $y + 8.88 = 9.99$

Subtract 8.88 from both sides $y = 1.11$

820. If you went swimming in that oil pool would you be more or less likely to sink than if you were in a pool filled with water. (Density of water is about 62 lbs./ft^3.)

    The buoyancy of a liquid is its upward force.
    buoyancy equals density times the volume displaced
    $b_{\text{in oil}} = dv$ becomes $b_{\text{in oil}} = 50v$
    $b_{\text{in water}} = dv$ becomes $b_{\text{in water}} = 62v$

You will be more buoyant in water and less likely to sink.

Complete Solutions and Answers | 840–841

840. Our duck decides to go out ice skating on the big lake. Is the distance he skates a discrete or a continuous variable?

It's a continuous variable. He might skate 4.9 feet or he might skate 388.7 feet.

Stop! I, your reader, object. Doesn't being a continuous variable mean that it can take on any value? I can name a distance that the duck will never skate.

Which is . . . ?

That duck will never skate 8,309,993,402,675,128,603 feet. Even if he skated his whole lifetime, he could never get that far.

Being a continuous variable does not mean that it can take on any value. It means that if, for example, it can take on the values of 4 and 7, then it can be any value between 4 and 7.

Ha! I don't believe you. Can you prove that the duck can skate exactly 5.398 feet?

That's easy. If he can skate 4 feet and if he can skate 7 feet, then we start him skating. At some point in time, he will have gone 4 feet. Later he will have gone 7 feet. At some instant in time he will have gone exactly 5.398 feet. If he didn't, then he would never get from 4 feet to 7 feet.

In contrast, the number of ducks I own is a discrete variable. I could own 4 ducks or I could own 7 ducks. But I'll never own exactly 5.398 ducks.

841. She painted right over the sign you had put on your door. The sign was 2.9 feet wide and 1.6 feet tall. What is the area of your sign?

$A_{rectangle} = lw$ becomes $\qquad A = (1.6)(2.9)$
Doing the arithmetic $\qquad\qquad\qquad\qquad A = 4.64$ square feet

Please recall that we are not using calculators in this book. This is one of your last real opportunities to learn your multiplication tables.

```
      2.9
    × 1.6
    -----
      174
       29
    -----
      464      4.64 square feet
```

| 842–844 | Complete Solutions and Answers |

842. In one hour she might steal two things from you. (1, 2). In 3 hours she might take 6 things. (3, 6). Graph these two points. Draw a line through them. Estimate how many things she'll steal in 4 hours.

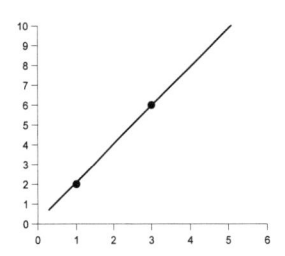

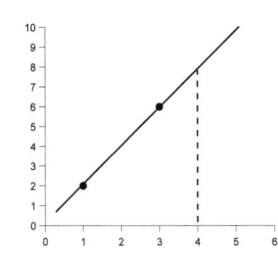

It looks like she'll steal 8 things in 4 hours.

843. He now weighs 242 pounds. It takes a hundred pounds of force to slide him. Find μ.

$$F = \mu N \quad \text{becomes} \quad 100 = 242\mu$$

Divide both sides by 242 and you get $\frac{100}{242} = \mu$

$\frac{100}{242}$ can be reduced to $\frac{50}{121}$

844. These four books plus the 3-pound onion weigh 28 pounds. All the Eldwood books weigh the same. How much does each Eldwood book weigh?

Let x = the weight of one Eldwood book.
Then 4x = the weight of all four books.
Then 4x + 3 = the weight of the books plus the weight of the onion
The whole thing weighs 28 pounds.

The equation becomes easy to write at this point. 4x + 3 = 28
    Subtract 3 from both sides   4x = 25
    Divide both sides by 4   x = 6.25 or 6¼
One Eldwood book weighs 6.25 pounds.

Complete Solutions and Answers   845

845. You lift that 5-ounce bar of chocolate from the table up to your nose. (That's an 18-inch trip.) There are few things that smell that good. How many ft-lbs of work did you do? 1 foot = 12 inches; 1 pound = 16 ounces

Since the answer is to be in foot-pounds, we first need to convert the ounces and inches into pounds and feet.

There are 12 inches in a foot. Then this fraction $\frac{1 \text{ foot}}{12 \text{ inches}}$ must be equal to one. (Any fraction where the top and bottom are equal is a fraction that is equal to one.)

Multiplying by one never changes the value of anything. In algebra we say that 1x is equal to x for any value of x.

So if I multiply 18 inches by $\frac{1 \text{ foot}}{12 \text{ inches}}$ I won't change the value.

$$\frac{18 \text{ inches}}{1} \times \frac{1 \text{ foot}}{12 \text{ inches}} = 1½ \text{ feet} \qquad (\text{the } inches \text{ canceled})$$

The $\frac{1 \text{ foot}}{12 \text{ inches}}$ is called a **conversion factor**. We are going to a lot of playing with conversion factors in this course. This is just the first taste. It sometimes takes a while to get used to them. Please take a peek at "conversion factors" in the index of *Life of Fred: Pre-Algebra 0 with Physics*. I count twenty-eight (28) pages in which we teach it over and over again. They will look easy by the time we're done.

Similarly, we convert 5 ounces into pounds.

$$\frac{5 \text{ ounces}}{1} \times \frac{1 \text{ pound}}{16 \text{ ounces}} = \frac{5}{16} \text{ pounds} \qquad (\text{the } ounces \text{ canceled})$$

The formula for work is W = Fd (work equals force times distance)

W = Fd becomes $\qquad\qquad$ W = $(\frac{5}{16})(1½)$

Doing the arithmetic $\qquad\qquad$ W = $\frac{15}{32}$ ft-lbs

$$\frac{5}{16} \times 1½ = \frac{5}{16} \times \frac{3}{2} = \frac{15}{32}$$

185

| 846–848 | Complete Solutions and Answers |

846. Your kid sister comes into your bedroom (without knocking). She takes three pencils out of your pencil jar. Does this make your jar more or less likely to start sliding?

    The coefficient of static friction, μ_s, depends only on the two surfaces that are in contact with each other. It does not depend on the normal force between those two surfaces. μ_s hasn't changed. rise/run hasn't changed. The jar is neither more nor less likely to start sliding.

847. She puts your pizza on top of the boxes. It does not slide.
    Then she gives it a little push and the pizza slides down the box on onto your lap. *This does not feel good.* Litotes means stating the opposite of what is true and putting a *not* in the sentence. For example,

Cheesecake with chocolate sauce is not a good breakfast.
Thirsty is not one of the days of the week.
Your father is not younger than you are.

    Does this show that $\mu_s < \mu_k$?
    It is harder to get something to start sliding than it is to keep it sliding. The coefficient of static friction, μ_s, is larger than the coefficient of kinetic (moving) friction, μ_k.

    $\mu_s < \mu_k$ is not true.

848. Graph (2, 6,), (3, 9), (4, 12).

I can draw a line through those points,
They are all in a row. (In geometry we will call those points collinear.)

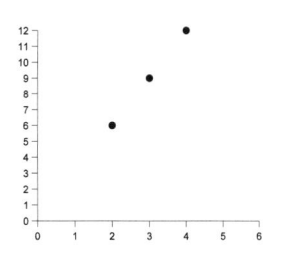

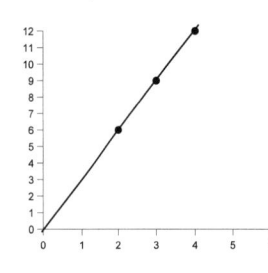

Mr. Hooke may have discovered Hooke's law (that the stretch is proportional to the weight on a spring) by graphing points and noticing that they are collinear. Graphing can help you see things.

186

Complete Solutions and Answers | 870–872

870. You have pizzas and strawberries next to your bed. There is only one more thing that you need. You install a giant barrel filled with chocolate milkshake. The faucet is 7 feet below the top of the barrel. What is the pressure at the faucet? (The density of good quality chocolate milkshake is 50 pounds/ft^3.)

p = dh (which was in the previous problem) becomes
p = 50 pounds/ft^3 × 7 ft = 350 pounds/ft^2

Is 350 lbs/ft^2 a lot of pressure?

Normal household city water pressure is typically in the range of 30–90 pounds per square inch. Let's say 50 psi.

How much is 50 psi in pounds/square foot?

$\dfrac{50 \text{ pounds/square inch}}{1}$ can be written as $\dfrac{50 \text{ pounds}}{1 \text{ square inch}}$ *
That will make the calculation easier to see.

$\dfrac{50 \text{ pounds}}{1 \text{ square inch}} \times \dfrac{144 \text{ square inches}}{1 \text{ square foot}}$

= 7,200 pounds/ft^2 is typical city water pressure.

Comparing this 7,200 lbs./ft^2 with the 350 lbs./ft^2 that is the pressure at your chocolate milkshake faucet means that your milkshake will flow leisurely** into your glass.

We also note that a good quality chocolate milkshake will flow more slowly than water because it is thicker. In physics we say that milkshakes are more **viscous** than water. They have a higher **viscosity**. (pronounced vis-COS-city)

872. What is the second coordinate of (20, 4)?

The second coordinate of (20, 4) is 4.

---

* Just as $\dfrac{3/4}{1} = \dfrac{3}{4}$

** The I before E except after C rule doesn't seem to apply to *leisurely*.

187

| 873 | Complete Solutions and Answers |

873. Kingie made $\frac{\text{rise}}{\text{run}}$ famous. If you tilt a flat surface upward until the block begins to slide, then the coefficient of static friction, μ_s, is equal to $\frac{\text{rise}}{\text{run}}$ In trig, we will define tan A to equal $\frac{\text{rise}}{\text{run}}$
So if ∠A is equal to 23°, then tan 23° = __?__

By eliminating all the extra words from the problem, we get

$\tan 23° = \frac{\text{rise}}{\text{run}} = \mu_s$ The answer is μ_s.

Now the fun part. In the old days, you needed to know F and N in order to find μ_s. If you didn't have a scale to measure force and weight, then Kingie's rise/run = μ_s would come in handy.

Suppose you didn't have a scale or a ruler. Suppose you only had a protractor (which is a piece of plastic used to measure angles.)

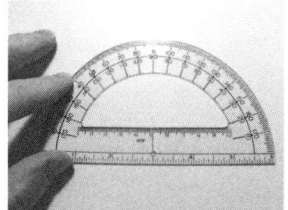

They cost about a dollar. You can get them at most office supply stores. Ask for one for your birthday.

Here's my protractor

You measure the angle and find . . . 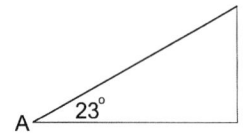
Now to find tan 23°.
You borrow your older sister's scientific calculator. You type in 2   3 and hit the tan key. Out pops 0.4244746 That's μ_s. You're done.

Intermission for calculators. Your calculator (which you are not to use for this course) is the basic one. It has +, −, ×, and ÷.

The second kind of calculator (the one your older sister has) is called a scientific calculator. It has keys like tan, sin, cos, log on it. It usually costs less than $10 in office supply stores.

This scientific calculator (which you will need for advanced algebra) is the last calculator you will ever need to get. You can use it all the way up through college calculus.

You will never need a graphing calculator, which costs about $100. I don't even own one, and I have a Ph.D. in math.

188

Complete Solutions and Answers — 874–876

874. If the kids had piled the second board on top of the first,

Would that change how much force he needed to apply?

No. The force needed to push or pull something is independent of the area of contact between the surfaces. μ didn't change. W didn't change. F = μW. F wouldn't change.

875. 1 cubic foot = 12^3 cubic inches.   1 ft^3 = 12^3 in^3
1 cubic yard = x^3 cubic inches. What does x equal?

    1 yard = 36 inches
    1 square yard = 36^2 square inches
    1 cubic yard = 36^3 cubic inches

876. If it takes Kingie 57 minutes to do an oil painting (such as "Swiss Mouse"), how long will it take him to do six of these paintings?

If you are not sure whether to add, subtract, multiply, or divide, restate the problem using very simple numbers. So, for example, if the problem was 2 minutes to do a painting, how long would it take to do 6 paintings? The answer is 12 minutes. You multiplied.

So we want 57 × 6.

```
     57
  ×   6
    342
```

It would take 342 minutes to do the six paintings.
If you wanted to convert that to hours, you divide by 60.

```
       5 R 42
  60)342
    -300
      42
```

It would take 5 hours and 42 minutes.

| 878–902 | Complete Solutions and Answers

878. 50¢ per swim   If 5,000 come and swim, how much will you make?

$$50¢ \times 5{,}000 = 250{,}000¢$$

Let's convert that to dollars.

The conversion factor is either $\dfrac{100¢}{\$1}$ or $\dfrac{\$1}{100¢}$

$$\dfrac{250{,}000¢}{1} \times \dfrac{\$1}{100¢} = \$2{,}500$$

900. The volume of a ball (called a **sphere** in mathematics) is $V_{sphere} = (4/3)\pi r^3$, where r = the radius of the sphere.
The radius of a sphere is the distance from the center to the surface. The diameter of the earth is roughly 8,000 miles. What is the volume of the earth? (Let $\pi = 3$ for this problem.)

$$V_{earth} = (4/3)\pi r^3 \text{ becomes} \quad V_{earth} = (4/3)(3)(4000^3)$$
$$= 4(4000^3)$$
$$= 4(64{,}000{,}000{,}000)$$
$$= 256{,}000{,}000{,}000 \text{ cubic miles}$$

901. If the radius of your stomach is 3", what is its volume? Let's suppose your stomach is in the shape of a sphere. $V_{sphere} = (4/3)\pi r^3$. (For this problem, let $\pi = 3$.)

$$V_{sphere} = (4/3)\pi r^3 \text{ becomes} \quad V_{stomach} = (4/3)(3)3^3$$
Doing the arithmetic $\quad V_{stomach} = 108 \text{ cubic inches}$

902. Did this change the force necessary to slide the bowl?

Yes. $F = \mu N$.

μ remained constant. If N changes, then F will change.

Complete Solutions and Answers | 903–904

903. You own 18.6 square miles in the shape of an ellipse. The area formula for an ellipse is $A = \pi ab$. a is 2 miles. How long is b?
For this problem we will let $\pi = 3.1$.

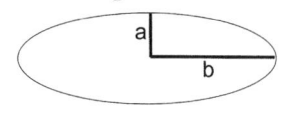

$A = \pi ab$ becomes	$18.6 = (3.1)(2)b$	
simplifying	$18.6 = 6.2b$	
dividing both sides by 6.2	$3 = b$	$6.2 \overline{)18.6}$

$$62. \overline{)186.} \quad \begin{array}{r} 3. \\ -186 \\ \hline 0 \end{array}$$

It's 3 miles from the center of your property to the east border.

904. The pizza is made out of rubber. Your kid sister switched the pizzas you ordered for a Jolly Joker Joking Pizza. When you bite down on it, it acts like a spring.

When you bite with 10 pounds of force, your teeth sink 0.4 inches into the pizza. If you bite with 6 pounds of force, how far will your teeth sink into it?

Hooke's Law is $F = kx$ where F is the force on the spring, where k is the constant for the spring, and where x is the amount of stretch (or compression) the spring experiences.

When you first bite down . . .

$F = kx$ becomes	$10 = k(0.4)$	
Put the number in front of the letter	$10 = 0.4k$	
Divide both sides by 0.4	$25 = k$	$0.4 \overline{)10.}$

$$4. \overline{)100.} \quad 25$$

Second, when you bite with 6 pounds of force . . .

$F = kx$ becomes	$6 = 25x$	
Divide both sides by 25	$0.24 = x$	$25 \overline{)6.00}$

$$\begin{array}{r} 0.24 \\ 25 \overline{)6.00} \\ -50 \\ \hline 100 \\ -100 \end{array}$$

Your teeth will sink 0.24 inches into the rubber pizza.

| 905–910 | Complete Solutions and Answers |

905. You order a crate of ducks. Two days later the 40-pound box of ducks is sitting on your living room floor. It takes 28 pounds of force to slide it into the dining room. Next week a 120-pound box of ducks arrives. How much force will be needed to slide it from the living room into the dining room?

    First of all, it doesn't matter that the second box has a greater area of contact with the living room floor than the smaller box. All that matters is the weights of the boxes.

    Since the second box is three times as heavy as the first box, it will take three times as much force to move it. 3 × 28 = 84 pounds.

910. If your barrel of chocolate milkshake were in the shape of a cylinder, with straight sides instead of curved sides, it might have these dimensions in feet.

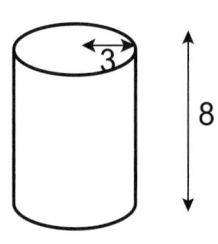

A) What is the area of the circle that has a radius equal to 3?
B) What is the volume of this cylinder? (Use Cavalieri's principle.)
C) What is the weight of the chocolate milkshake in this container? Hint: We know the density from the previous problem. We know the formula from the first problem in the *Your Turn to Play*.

A) $A_{circle} = \pi r^2$ becomes $A = \pi 3^2 = 9\pi$ square feet.

B) Each of the "poker chips" has an area of 9π square feet. Cavalier's principle says that the volume will equal the area of a poker chips times the height. In this case, $V_{milkshake\ cylinder} = 9\pi \times 8 = 72\pi$ ft^2.

C) The formula from #1 in the *Your Turn to Play* is dv = w.

    This becomes (50 lbs./ft^3)(72π ft^3) = 3,600π lbs.

That is roughly equal to (3,600)(3) = 10,800 lbs.

My truck weighs 4,049 lbs.

Complete Solutions and Answers | 918–919

918. An alligator blown 50 feet into the air. What form of energy did you change into what other form?

     Exploding dynamite changes the chemical energy stored in the explosive into sound/heat/light. Heat turns into motion as it expands the air.

     The motion of the air on the tummy of the alligator pushes it up into the air. At a height of 50 feet, all the motion is converted into the energy of height.

919. You attach that 18-pound alligator to the ceiling with a spring. It stretches the spring 1.2 meters. You then take the big rubber ball, which weighs 6 pounds) and balance it on the alligator's nose. How long will the spring be stretched now?

    Hooke's law is $F = kx$

          where F is the force on the spring and x is the distance the spring stretches.

Eighteen pounds stretches the spring 1.2 meters. $18 = k(1.2)$

Put the number in front of the letter $18 = 1.2k$

Divide both sides by 1.2 $\dfrac{18}{1.2} = k$

We now know the value of the spring constant, k.

We want to know how far the spring stretches when 24 pounds (= 18 + 6) is attached.

Translation: We want to know x when F = 24.

F = kx becomes $24 = \dfrac{18}{1.2} x$

Divide both sides by $\dfrac{18}{1.2}$ $24 \div \dfrac{18}{1.2} = x$

Doing the arithmetic $1.6 = x$

The spring will stretch 1.6 meters.

$$24 \div \dfrac{18}{1.2}$$

$$\dfrac{24}{1} \times \dfrac{1.2}{18}$$

$$\dfrac{\cancel{24}^{4}}{1} \times \dfrac{1.2}{\cancel{18}_{3}}$$

920–921 Complete Solutions and Answers

920. How many pounds does each of those 0.6 kg strawberries weigh? Use a conversion factor. 1 kg ≈ 2.2 lbs.

The conversion factor will be either $\frac{2.2 \text{ lbs.}}{1 \text{ kg}}$ or $\frac{1 \text{ kg}}{2.2 \text{ lbs.}}$

$\frac{0.6 \text{ kg}}{1} \times \frac{2.2 \text{ lbs.}}{1 \text{ kg}} = 1.32$ lbs.   Those are big strawberries.

921. "If we were now at our vacation home that is near the pass that goes into Yosemite (elevation 10,000 feet), then you could use this 34-foot straw to suck up Sluice." Does this sound true to you?

Your first reaction was *What vacation home? You have never mentioned this to me! Did you just buy one?* Then you realized that you can tell three-year-olds a lot of counterfactual things* such as that reindeer can fly, and they will believe you without questioning.

But even three-year-olds know that the pressure of Sluice in a straw is equal to the density of the Sluice times the height of the liquid. That's just p = dh.

And three-year-olds know that the pressure exerted by the column of Sluice is equal to the pressure on the surface of the Sluice Lake. It's not a vacuum that sucks the Sluice upward. At higher altitude, the pressure is less, and the column of Sluice will be lower.

What your mother should have said is, "If we were now at our vacation home that located on the planet Jupiter (where the gravity is much more than here on earth), then you could use this 34-foot straw to suck up Sluice."

---

* counterfactual = not according to the facts = lies

Complete Solutions and Answers | 922–923

922. Work is a transfer of energy. Into what form of energy is she transferring the energy as she slows you down?
(The nine forms of energy are chemical, electrical, heat, height, light, motion, nuclear, sound, and spring.)

    Her shoes scraping along the sidewalk will generate heat.

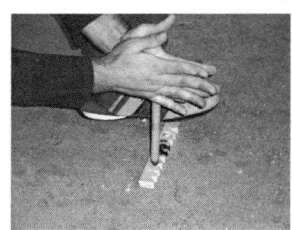

Rubbing two pieces of wood together is one way of creating fire. Friction is turned into heat. On the Internet you can find detailed instructions. Search for something like `fire rubbing two pieces of wood`.

923. This floating alligator displaces what volume of the Sluice?
The alligator weighs 18 pounds, and the density of Sluice is 0.8 lbs./in³.

    Since it is floating, its weight (18 pounds) is equal to its buoyancy (18 pounds).

    Buoyancy equals the density of the liquid times the volume of the liquid displaced.

Buoyancy = dv becomes	$18 = (0.8)v$
Dividing both sides by 0.8	$\dfrac{18}{0.8} = v$
Doing the arithmetic	$22.5 = v$

The alligator displaces 22.5 cubic inches of Sluice.

Just for fun, I'm going to redo these three lines and include all the units. This will show that v has the units of cubic inches.

Buoyancy = dv becomes	$18 \text{ lbs.} = (0.8 \text{ lbs./in}^3)v$
Dividing both sides by 0.8 lbs./in³	$\dfrac{18 \text{ lbs.}}{0.8 \text{ lbs./in}^3} = v$
Doing the arithmetic	$\dfrac{18 \text{ lbs.}}{1} \div \dfrac{0.8 \text{ lbs.}}{\text{in}^3}$

| 924-925 | Complete Solutions and Answers |

$$\frac{18 \text{ lbs.}}{1} \times \frac{\text{in}^3}{0.8 \text{ lbs.}}$$

$$\frac{18 \cancel{\text{lbs.}}}{1} \times \frac{\text{in}^3}{0.8 \cancel{\text{lbs.}}}$$

22.5 in³

924. Did this change your attitude toward your kid sister?

No. You still love her. After all, she is only 3 and still needs to learn about sharing. In the future you might want to take a bit of the chicken before you pass the bowl to your kid sister.

925. You put the half-eaten bar of chocolate into your microwave oven and set it for 40 seconds. It is turned into a warm, gooey, delicious puddle. Energy has been converted from what form into what form?

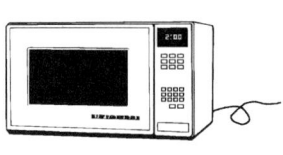

Microwave ovens operate on electricity. (At least, I've never seen one operate on gas.)
The electricity is transformed into heat energy.*
The result is tasty chocolate "soup."
(Cooking hint: Put the bar of chocolate in a bowl before you microwave it. Otherwise, you'll have a real mess.)

---

* There are several intermediate steps between the electrical energy and the heat. You don't notice them when you are using a microwave oven.

The electricity is first changed into microwaves. The microwaves agitate the water in the food. That creates friction. And friction creates heat.

Complete Solutions and Answers 926–928

926. If we make the rope one hundred times as long, how will that affect the period?

    That's the opposite of the previous question.

    The period is proportional to $\sqrt{\text{length}}$.

    If I make the rope one hundred times as long, the period will be ten times as long. $\sqrt{100} = 10$

927. With her little chubby hands, she moves the jar a little bit when she is stealing your pencils. Does this make the jar more likely to start sliding?

    The coefficient of kinetic (moving) friction, μ_k, is less than the coefficient of static friction, μ_s. Translation: It is always easier to push something that is moving than it is to get it started.

    Since she has started the jar moving, there is an increased chance that it will stay moving.

    It moved. It fell. It broke.

928. $3x - 5 = 2x$ and $30 = 5y + 9$

    Is it true that $x < y$?

    In order to find out if $x < y$, we first have to find out what x is and what y is. (Pretty logical, isn't it?)

We start with	$3x - 5 = 2x$
Add 5 to both sides	$3x = 2x + 5$
Subtract 2x from both sides	$x = 5$

We start with	$30 = 5y + 9$
Subtract 9 from both sides	$21 = 5y$
Divide both sides by 5	$4.2 = y$

$$5 \overline{)21.0} \quad = 4.2$$

Now that we know that x is 5 and y is 4.2, we can say that $x < y$ is not true.

929–951 Complete Solutions and Answers

929. We want to solve $3x + 5 = 32$.
The first step is to subtract 5 from both sides.

> We start with $\qquad 3x + 5 = 32$
> Subtract 5 from both sides $\qquad 3x = 27$

At this point it becomes "old stuff" that you already know.

> Divide both sides by 3 $\qquad x = 9$

Hey! I, your reader, have a question. When you start with $3x + 5 = 32$, why can't you first divide by 3?

> You could . . . but it would turn into a mess.

If you divide $3x + 5 = 32$ by 3 you get $3x/3 + 5/3 = 32/3$.
Tons of examples:

> If you have $70 = 11w + 4$, you first subtract 3 from both sides.
> $66 = 11w$

> If you have $19 + 2x = 33$, you first subtract 19 from both sides.
> $2x = 14$

> If you have $1,000,000 = 999,994 + 6y$, you first subtract 999,994 from both sides.
> $6 = 6y$

951. You decide to sing your *Fishy Goodbye Song* as you walk toward the wastebasket. You start with the lung muscles and wind up with song. Describe how energy is converted among the nine forms of energy: *motion/heat/light/sound/electrical/height/nuclear/spring/chemical*.

> To make your muscles move, you convert *chemical* energy into *motion*.
> The *motion* of your muscles is converted into the *motion* of air through your vocal chords.
> The *motion* of air through your vocal chords is converted into *sound*.

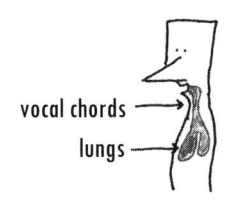

Complete Solutions and Answers | 952–954

952. You walk over to the door, and she suddenly opens the door. The doorknob hits you with 28 pounds of force. Do you wish that the area of that doorknob was larger or smaller than it is now?

It is the pressure that the doorknob makes that is important. Pressure is defined as Force/Area.

The force is 28 pounds. Pressure = $\dfrac{28 \text{ pounds}}{\text{Area}}$

The smaller the area is, the larger $\dfrac{28 \text{ pounds}}{\text{Area}}$ will be.

If the doorknob were the size of the whole door, then it wouldn't hurt much. If the doorknob had the area of a needle point, it would go right through you. That's why pillow fighting is fun and not dangerous.

953. For the last three years every time you do something that your kid sister doesn't like, she cries. You figure that when you get to the garbage can, she'll be crying when you throw the glob away. Is this an example of inductive or deductive reasoning?

Your conclusion is based on experiments/trials/observations—the three things that scientists use to form their hunches, conjectures, theories, and laws. They use inductive reasoning, and their conclusions are never 100% certain.

Who knows? When you throw that silly popped glob in the garbage can, your kid sister may smile and thank you for helping her clean up that mess.

954. The rise/run is 8/20, so $\mu_s = \dfrac{8}{20}$

Doing the arithmetic, $\mu_s = 0.4$ or $\dfrac{2}{5}$

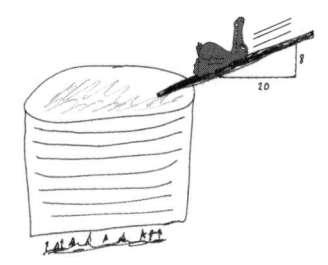

199

| 963–976 | **Complete Solutions and Answers** |

963. Light can travel one meter in $\frac{1}{299{,}792{,}458}$ of a second.

That's the same thing as saying it can travel 299,792,458 meter in one second. (It's fast.) How far can it travel in 4 seconds?

It can go four times as far. 4 × 299,792,458

$$\begin{array}{r} 299{,}792{,}458 \\ \times 4 \\ \hline 1{,}199{,}169{,}832 \end{array}$$

It can travel one billion, one hundred ninety-nine million, one hundred sixty-nine thousand, eight hundred thirty-two meters in four seconds.

970. A 16" pizza is a large pizza in most pizza restaurants. The 16" is the measure of the diameter. What is the area of that pizza?

In problem #255 in Chapter 15 we found that $A_{circle} = \pi r^2$. A 16" diameter means that the radius is 8 inches. $A_{circle} = \pi r^2 = \pi 8^2 = 64\pi$ square inches.

Discussion: In the previous problem (#718) we found that the area of your snack pizza was approximately 230.4 square inches.

This 16" pizza has area of 64π which is approximately 64 × 3.14 = 200.96 ≑ 201 square inches. Your snack pizza is larger than a traditional 16-inch pizza.

976. Solve   3w + 61 = 15w + 1
    Subtract 1 from both sides   3w + 60 = 15w
    Subtract 3w from both sides   60 = 12w
    Divide both sides by 12   5 = w

Complete Solutions and Answers | 981–986

981. The ellipse measured 6 feet across at its widest and 4 feet at its narrowest. What was the area of this puddle? Use $\pi = 3$ for this problem.

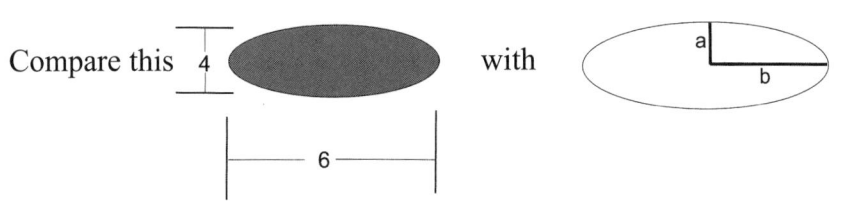

Compare this with

and you can see that a = 2 and b = 3. $A = \pi ab$ becomes $A = (3)(2)(3)$

   The area of the puddle is 18 square feet.

986. A point has no dimension. A segment has one dimension. A square has two dimensions. A cube has three dimensions. Four dimensions?

   One of the reasons math is exciting is that we do stuff that boring people find impossible to accept. In algebra, for example, we will show you how to compute $2^{1.5}$ and 2^{-3}. In later math we will count the number of natural numbers, $\{1, 2, 3, 4, \ldots\}$ and tell you how many there are.*

   And what comes after point, segment, square, and cube? It is the tesseract. **Hey! Stop! This is not possible. My teachers, parents, guardians, jailers, newscasters (choose one) have told me I live in a three-dimensional world. You can't have a four-dimensional "cube."**

   No? You can't have one because *you* think you live in a three-dimensional world. That doesn't mean *I* have to live in your little world. In *Life of Fred: Geometry Expanded Edition* I will draw a four-dimensional "cube"—a tesseract. And a cube in five dimensions (the hypertesseract). And we will compute the number of edges that a "cube" in the 14th dimension has. (If you need to know now, it's 114,688.)

   Everyone who has held a die (that's the singular of dice) in his hand knows that it has six faces. The "cube" in the 14th dimension has 372,736 faces and we will show you how to compute that number.

---

* And we won't just say "infinity." **I, your reader, demand to know right now. How many natural numbers are there?** Okay. I'll tell you, but you are not used to looking at numbers beyond the finite. There are $\aleph_0$ of them. And there are numbers bigger than $\aleph_0$. If you want an example, $\aleph_1$. And $\aleph_1 \neq \aleph_0$. That's because $\aleph_1 > \aleph_0$. And we *prove* that is true. There is no need to accept that on faith.

990 Complete Solutions and Answers

990. To solve $13 = x + 6$ you subtract 6 from both sides and get $7 = x$.

Solve

$$18 = x + 2$$
Subtract 2 from each side $\quad 16 = x$

$$200 = y + 50$$
Subtract 50 from each side $\quad 150 = y$

$$w + 4 = 11$$
Subtract 4 from each side $\quad w = 7$

$$6 + z = 60$$
Subtract 6 from each side $\quad z = 54$

$$33 = 30 + x$$
Subtract 30 from each side $\quad 3 = x$

Index

algebra
- #121. 62
- #124. 19
- #160. 16
- #161. 49
- #258. 63
- #316. 62
- #369. 27
- #392. 49
- #487. 70
- #492. 17
- #605. 63
- #615. 17
- #619. 77
- #620. 81
- #641. 39
- #661. 88
- #797. 17
- #817. 31
- #819. 51
- #928. 85
- #929. 51
- #976. 79
- #990. 46

area and volume
- #162. 60
- #190. 66
- #211. 19
- #260. 41
- #288. 22
- #311. 9
- #486. 19
- #488. 60
- #490. 27
- #505. 87
- #506. 69
- #709. 49
- #710. 65
- #740. 89
- #760. 39
- #841. 39
- #900. 61
- #901. 62
- #903. 25
- #910. 63
- #974. 49
- #981. 29

buoyancy
- #125. 70
- #190. 66
- #213. 68
- #340. 66
- #426. 91
- #450. 64
- #500. 68
- #573. 64
- #626. 70
- #676. 64
- #820. 89
- #923. 79

Index

Cavalieri's principle
- #535. 62
- #589. 91
- #674. 62
- #910. 63

conversion factor
- #106. 47
- #210. 65
- #212. 52
- #364. 48
- #366. 59
- #393. 90
- #588. 84
- #679. 85
- #680. 87
- #718. 49
- #756. 57
- #761. 79
- #762. 50
- #876. 32
- #920. 59
- #963. 15

degrees—geometry
- #120. 61
- #200. 40
- #314. 61
- #499. 63
- #596. 40
- #659. 61
- #660. 40
- #720. 40
- #873. 45

density
- #614. 68
- #740. 89
- #758. 68

discrete and continuous variables
- #404. 10
- #560. 10
- #643. 25
- #717. 13
- #763. 25
- #840. 11

distance—rate—time
- #155. 10
- #818. 87

energy in 9 forms
- #156. 44
- #158. 30
- #167. 78
- #289. 44
- #367. 30
- #428. 30
- #445. 91
- #495. 65
- #564. 49
- #918. 42
- #922. 55
- #951. 67

Index

$F = \mu_k N$—kinetic (moving) friction
- #209. 51
- #215. 69
- #313. 55
- #383. 41
- #405. 67
- #449. 51
- #454. 54
- #489. 41
- #594. 35
- #726. 85
- #750. 33
- #847. 65
- #927. 83

$F = \mu N$
- #192. 18
- #214. 16
- #262. 18
- #318. 13
- #395. 16
- #455. 14
- #457. 22
- #502. 13
- #555. 22
- #584. 14
- #601. 22
- #657. 15
- #711. 22
- #744. 18
- #805. 15
- #843. 17
- #874. 15
- #902. 18
- #905. 11

$F = \mu_s N$—static friction
- #126. 32
- #194. 33
- #215. 69
- #263. 33
- #317. 28
- #319. 34
- #360. 33
- #405. 67
- #572. 28
- #594. 35
- #723. 36
- #724. 81
- #745. 83
- #847. 65
- #873. 45
- #927. 83
- #954. 57

friction
- #422. 9
- #453. 33
- #494. 11
- #640. 11
- #715. 11
- #721. 71
- #809. 55
- #846. 83

Index

graphing and ordered pairs
- #198. 56
- #266. 20
- #287. 45
- #451. 20
- #534. 45
- #595. 20
- #716. 20
- #725. 77
- #842. 29
- #848. 85
- #872. 21

Hooke's law for springs—$F = kx$
- #189. 26
- #315. 25
- #352. 26
- #498. 25
- #536. 26
- #673. 26
- #719. 71
- #904. 37
- #919. 31

inductive and deductive reasoning
- #344. 73-75
- #420. 23
- #570. 24
- #659. 61
- #722. 31
- #953. 33

metric system
- #105. 23
- #119. 14
- #157. 50
- #181. 12
- #212. 52
- #267. 42
- #286. 60
- #485. 52
- #491. 50
- #544. 39
- #571. 60
- #639. 50
- #714. 42
- #762. 50

numerals
- #208. 9
- #456. 39

Ohm's law—V, I, and R
- #259. 86
- #281. 90
- #393. 90
- #424. 88
- #661. 88

parallel circuit
- #115. 90
- #201. 90

Index

pendulum and period
- #251. 12
- #280. 13
- #597. 12
- #755. 12
- #799. 43

percents
- #164. 80
- #261. 32
- #610. 59
- #751. 32
- #764. 87
- #798. 51

pressure
- #165. 69
- #284. 58
- #458. 76
- #600. 58
- #644. 79
- #710. 65
- #713. 63
- #765. 89
- #870. 63
- #952. 83

proportional
- #677. 10

schematic diagrams
- #110. 82
- #127. 88
- #166. 84
- #217. 82
- #240. 84
- #265. 86
- #270. 88
- #459. 87
- #461. 82
- #504. 83
- #507. 84

sketching W, N, and F
- #285. 38
- #658. 83

speed of light—c
- #105. 23
- #119. 14
- #365. 10
- #712. 45
- #963. 15

vacuum
- #368. 72
- #398. 72
- #458. 76
- #759. 72
- #921. 77

word problems
- #114. 54
- #120. 61
- #193. 73
- #314. 61
- #421. 56
- #496. 67
- #598. 56
- #619. 77

Index

    #681. 89
    #754. 59
    #844. 57

work
    #123. 43
    #312. 43
    #333. 81
    #356. 65
    #460. 81
    #491. 50
    #497. 43
    #642. 67
    #678. 71
    #726. 85
    #757. 67
    #845. 42